Mary F. (Mary Foote) Henderson

Diet for the Sick

A Treatise on the Values of Foods, Their Application to Special Conditions of Health and Disease, And on the Best Methods of Their Preparation

Mary F. (Mary Foote) Henderson

Diet for the Sick
A Treatise on the Values of Foods, Their Application to Special Conditions of Health and Disease, And on the Best Methods of Their Preparation

ISBN/EAN: 9783744646055

Printed in Europe, USA, Canada, Australia, Japan

Cover: Foto ©berggeist007 / pixelio.de

More available books at **www.hansebooks.com**

DIET FOR THE SICK

A TREATISE ON THE VALUES OF FOODS, THEIR APPLICATION
TO SPECIAL CONDITIONS OF HEALTH AND DISEASE, AND
ON THE BEST METHODS OF THEIR PREPARATION

BY

Mrs. MARY F. HENDERSON
AUTHOR OF "PRACTICAL COOKING AND DINNER GIVING"

ILLUSTRATED

"*Man kills himself, rather than dies*"

NEW YORK
HARPER & BROTHERS, FRANKLIN SQUARE

TO

Dr. EDWARD BAYARD, of New York
AND
Dr. T. GRISWOLD COMSTOCK, of St. Louis

THE FORMER THE LIFE-LONG FRIEND OF MY FATHER, AND BOTH
EMINENT PHYSICIANS FOR WHOM I HAVE UNBOUNDED
RESPECT AND ESTEEM, THIS BOOK

Is Gratefully Dedicated

PREFACE.

An English author says, "The doctor, unsupported by the cook's material aid, and the cook, unguided by the doctor's knowledge, are two powerful agents, half of whose strength is paralyzed or misdirected."

There are many valuable books published on "dietetics," which give at great length the chemical analyses of foods, their flesh-building and heat-producing values, etc. Many books are also published containing receipts for the sick.

But the wise suggestions contained in the quotation have been too little heeded. The doctors have not been cooks. The cooks have not been doctors.

The author, although neither a doctor nor a cook (as high art can reach in that direction), still concluded to modestly venture into the domains of both, with the view of studying the relations of foods to health and disease, and also the best modes of preparing them for use.

The subject is an important one. A proper dietary is surely as essential to the recovery of an invalid as medicine; and yet it will be observed that medical works give a thousand pages to medicinal therapeutics to one of dietetics.

A physician at the head of one of our medical colleges writes: "An experience of ten years in examining medical students, reveals a tendency in them to consider themselves 'medicine men,' as is shown by the undue importance they attach to the pills, powders, and potions they are prepared to give. They ignore the great fact that upon a carefully selected diet the patient must depend for the natural strength which is to be the chief reliance in stemming the adverse tide."

The professor might have added more; for the treatment of most chronic complaints is chiefly dietetical and hygienic, rather than medicinal.

The science is comparatively new which treats of the chemical composition, preparation, and physiological effects of foods. May the time come when nothing in the way of dietetical treatment need be left to the fickle appetites of patients, nothing to the judgments of anxious friends, and nothing to untrained cooks!

If this hand-book — intended to be small, though the subject is large — can be of use in the direction just indicated, it will give great pleasure and satisfaction to

THE AUTHOR.

ST. LOUIS, *March*, 1885.

CONTENTS.

REMARKS ABOUT BEVERAGES AND FOODS ....Page 1
    Tea... 1
    Coffee 5
    Cocoa... 5
    Chocolate... 5
    Iced Water and Iced Tea 6
    Spirituous Liquors... 8
    Malt Extract... 9
    Milk.. 10
    Buttermilk 13
    Whey... 13
    Animal Foods.. 13
    Salted Meats.. 16
    Fish.. 16
    Oysters.. 16
    Fat.. 18
    Eggs.. 20
    Rice.. 20
    Corn-starch and Arrow-root 21
    Sago and Tapioca.. 21
    Beans and Pease 21
    Gelatine... 21
    Tomatoes.. 22
    Fruits, Grapes, Bananas, etc..................................... 22
    Sea-moss Farine and Sea Moss.................................... 25
THE NEW HEALTH FOODS AND OTHER GRAIN PREPARATIONS... 26
KOUMISS... 31
ARTIFICIAL DIGESTION BY MEANS OF PANCREATIC FERMENTS.. 39
GRAPE JUICE.. 44

Contents.

THE HOT-WATER CURE.................................Page 47
DIET IN DIFFERENT DISEASES, ETC..................... 49
    Diet for Babies.. 49
    Dyspepsia.. 55
    Diarrhœa.. 61
    Dysentery.. 62
    Cholera.. 62
    Fevers... 64
    Typhoid Fever.. 66
    Gout and Rheumatism................................. 67
    Bright's Disease...................................... 69
    Diabetes.. 71
    Consumption... 73
    Scrofula.. 75
    Rickets... 75
    Diphtheria... 76
    Gastritis... 76
    Corpulency.. 77
    Colds and Catarrhs................................. 225
SOMETHING ABOUT LONGEVITY........................... 80
UTENSILS... 85
RECEIPTS FOR THE SICK AND CONVALESCENT............. 89
    Drinks.. 89
    Beef Teas and Broths................................ 100
    Gruels... 106
    Breads and other Grain Preparations................ 113
    Receipts for Gluten................................. 130
    Vegetables.. 134
    Little Dishes....................................... 139
    Some Cream Soups.................................... 158
    Other Soups.. 164
    Dishes of Rice..................................... 169
    Creams and Fruits.................................. 174
    Custards.. 180
    Jellies... 184
    Puddings.. 190
BILLS OF FARE FOR CONVALESCENTS.................... 194

APPENDIX... 199
    EFFECTS OF TEA AND COFFEE.
        Extract from article by M. Mattieu Williams........ 199

Contents.

APPENDIX (*continued*).

INFLUENCE OF ALCOHOLIC LIQUORS.
  Remarks on the subject by Prof. Edward L. Youmans and others..Page 205

TENDENCY OF COMMON WHEAT FLOUR TO PRODUCE BRIGHT'S DISEASE, DIABETES, ETC.................................. 207

SOMETHING ABOUT KOUMISS.
  Extract from an article by Dr. E. F. Brush of New York, in the *Medical Record*................................ 208

MORE ABOUT KOUMISS.
  By Dr. T. Griswold Comstock of St. Louis............... 211

THE DIGESTIVE FERMENTS.
  Extracts from a book on the subject by Dr. William Roberts of Manchester, England........................ 212

PANCREATIC EMULSION OF FATS.
  Extract from a work on "Loss of Weight, Blood Spitting, and Lung Disease," by Dr. Horace Dobell............ 213

FOOD FOR INFANTS.
  Remarks by Dr. Eustace Smith, Physician to the King of the Belgians.. 215

FEEDING THE BABY.
  Remarks by Dr. E. C. Page, in his book "How to Feed the Baby"... 221

DIET FOR TYPHOID FEVER.
  Extracts from an address on the "Treatment of Typhoid Fever," by Sir William Jenner...................... 223

COLDS AND CATARRHS.
  Extract from an article by Dr. Felix Oswald, published in the "Popular Science Monthly".................... 225

MORE ABOUT THE PANCREATIC EXTRACT..................... 228

ALPHABETICAL INDEX.................................. 231

DIET FOR THE SICK.

REMARKS ABOUT BEVERAGES AND FOODS.

Tea.

This article contains an astringent matter, tannin (constituting from eighteen to twenty-five per cent. of the whole), a volatile oil yielding the aroma, and theine. Authorities differ regarding the effects of tea upon the system. Liebig claims that theine and caffeine, in some way not satisfactorily explained, prevent or suspend the waste of tissue. As nitrogenous foods are necessary to supply or reinforce the tissues, he thinks that whatever prevents the waste of tissue takes the place of such foods, and, *pro tanto*, dispenses with the necessity of their use. This theory is now substantially exploded. More thorough investigation, supplemented by careful experiments, has demonstrated almost beyond question that the waste of tissue is not, in fact, prevented by tea or coffee. The essential principles of both are stimulating in their effects. They increase the action of the heart and the arteries, and furnish, like alcohol, a transient increase of vital energy. These stimulating effects have, by Liebig and some others, been accepted as the prolongation or maintenance of healthful strength and vital force. These effects are deceptive. The advantages are as unreal as those sometimes supposed to spring from the use of alcoholic liquors. In truth, if alcohol were taken in small enough quantities to produce an effect no more stimulating than the use of tea, it would be less injurious to the system, from the fact that tea is mixed with the food, adulterating it with tannin, which is not contained in alcohol.

Dr. Bellows considers Liebig's theory fallacious, and attributes the benefits of tea rather to its osmazone (the flavoring principle). He says that food is more digestible and assimilable when it is taken with gustatory pleasure. The aromatic principle of tea commends it to the taste. He instances an experiment on a dog that was shut up and given good natural food containing all the needed elements except osmazone; *i. e.*, the food was cooked and recooked until all flavor and odor were lost. The dog finally refused to eat and pined away. It may be possible that the overcooking renders the fibres and other elements unfit for digestion, making them tough and depriving them in some degree of nutritive power. The dog's food, in the case named, was, perhaps, little better for dietetic purposes than so much wood. Flavor, or the sense of taste, is possibly an index by which nature, unperverted, determines the proper food to be taken into the stomach at any given time.

Professor Lehman also believed that tea and coffee lessen the waste of the body. Dr. Edward Smith believes to the contrary. He says (in "Foods"), "I performed a very extended series of experiments on myself and others, which proved that tea excites vital action, and is practically a respiratory stimulant. . . . In reference to nutrition, tea increases waste, since it promotes the transformation of food without supplying nutriment, and increases the loss of heat without supplying food." Tea, therefore, he thinks, should only be taken after a full meal, unless the system be at all times replete with nutritive material.

Pavy says: "The phenomena produced when tea is consumed in a strong state, and to a hurtful extent, show that it is capable of acting in a powerful manner upon the nervous system. Nervous agitation, muscular tremors, a sense of prostration, and palpitation constitute effects often seen. It also possesses direct irritatant properties which lead to the production of abdominal pains and nausea, and by the astringent matter it contains it diminishes the action of the bowels."

Some authorities, indeed, go so far as to say that tea is a most potent destroyer of the digestive powers.

There are probably some good results from drinking tea and coffee, viz.: the water used is purified by boiling, the liquid is generally taken in a warm state, and the warmth of the water tends to aid digestion. When milk or cream is used a valuable nutritive aliment is added which might not otherwise be taken; and possibly, as one authority remarks, the use of tea or coffee in many cases furnishes a sufficient stimulus to protect against indulgence in drinks of a still more injurious character. In other words, it is thought that dying of tea and coffee is more gradual than dying of whiskey and brandy.

It is now generally conceded that the effect of the active principle in tea and coffee is more or less injurious to the nervous system, and the tannin contained in them acts as a constant irritant to the stomach, presenting a formidable obstacle to digestion. Slavery of body and mind to any unnatural stimulant is unfortunate, whether that stimulant be tea or coffee, alcoholic drinks or opium—all more or less beneficial as remedial agents and injurious as constant beverages.

The feeling of health and strength which makes it a luxury to live, the exhilarating sense of self-command which makes work a pleasure and success a certainty, that happy buoyancy of spirit which comes only from the taking of wholesome and assimilable food, cannot be properly appreciated by those who depend upon the ephemeral effects of stimulants.

If a stimulating drink is desired, nothing is more wholesome than koumiss before it becomes too acid. Chocolate (alkathrepta made without vanilla) furnishes another nourishing, although hardly a stimulating, drink.

Several substitutes for coffee have been tried, such as chiccory, roasted beans, pease, etc. Probably the best substitute is the cereal coffee prepared by the Health Food Company. It is made of the entire barley grain and the gluten of wheat. It is of nutritive value, and has a pleasant flavor resembling coffee. One tires of it, however, after a short time. The most pleasant and innocent of drinks for a constant beverage is one at the mention of which the reader may smile incredulously. But, let him first try it. I call it hot-water tea.* It consists simply of boiling water, with cream and sugar added, and is served in a teacup. The temperature of the boiling water should be properly reduced by the addition of the cold cream. As soon as hot-water tea is given a fair trial, it will be discovered that it is chiefly the warmth of the beverage that is desired; also that, with a bit of imagination, hot-water tea will soon seem to possess all the flavor of the genuine English Breakfast or Hyson, the Government Java or the Mocha infusions. Yet some persons are very *difficile*, and have no imagination.

In closing this article, I would add that green tea is more objectionable than black tea. It contains a third more tannin, and often a deleterious coloring matter

---

* The author has since heard that this beverage is mentioned in the menus of some New York hotels as "cambric tea."

(Prussian blue mixed with gypsum and indigo). In the preparation of tea it should never be allowed to boil and steep. Boiling water should be poured upon the leaves, and the infusion used in a very few minutes afterwards. The tea leaves should never be used a second time. When tea is boiled, tannin is extracted in undue quantities, and the volatile osmazone is driven off.*

COFFEE.

Coffee is heating and stimulating, and is serviceable in giving warmth to the body under exposure to cold. Taken in immoderate quantities it induces feverishness, tremor, palpitation, anxiety, and deranged vision. It contains less tannin than tea, and is probably less injurious to the digestive powers.

COCOA AND CHOCOLATE.

Cocoa is the name of the seed or bean of the cacao-tree, ground into a powder, and moulded into cakes. When it is flavored with vanilla and mixed with sugar it is called chocolate. For the invalid, chocolate should be avoided on account of the vanilla. Cocoa or alkathrepta (a quite pure pharmaceutical preparation) should be substituted for it. Cocoa differs from tea and coffee in that it possesses little or no tannin, or other of their deleterious constituents. It contains a large percentage of fatty and albuminous (muscle-making) matter, with about four per cent. of phosphates, and is supplied with all the requisite elements of food for sustaining life. It possesses the stimulating effect of tea and coffee, though in a very mild degree. Pavy says: "Containing, as cocoa does, twice as much fatty matter as wheaten flour, with a notable quantity of starch, and an agreeable aro-

---

* For further remarks about tea and coffee see Appendix, page 199.

ma to tempt the palate, it must be a valuable alimentary material. Chocolate taken with milk and bread will suffice for a good repast." The nutritive elements of cocoa are so concentrated, and it is so rich in oily matter, that it should only be freely taken by convalescents and persons in active life. In Solis's "Conquest of Mexico" it is said that the Spanish conquerors did not fail to record their appreciation of the flavor and nutritive qualities of chocolate, a single cup of it being enough, in their estimation, to sustain a man through a day's march. The cups were probably large.

Chocolate is frequently adulterated with starch, suet, and coloring matters. Venetian red, umber, annatto, and, in some instances, the highly poisonous metallic salts of cinnabar and red lead are employed.

The chocolate in common use is, therefore, of very uncertain composition. According to Dr. Hassall the doubtful article composes half of what is sold in England.

Iced Water and Iced Tea.

The digestive agents are very sensitive to temperature, the process of digestion being arrested by a temperature either too hot or too cold. This is practically tested by experimenting with the receipts given in this book, where the pancreatic extract is employed.

Water, to be refreshing and wholesome, should not, when drunk, be above the usual temperature of fresh spring or well water. The habitual use of iced water by Americans is certainly attended with great injury; and undoubtedly this lavish use of it and hot breads is the chief cause of the national disease—dyspepsia. A waiter's first duty at an American hotel is to place before each guest a goblet full of cracked ice, and the crevices are then filled with water before he takes an order for something else to go with it.

An acquaintance of the writer, in Missouri, who was blessed with a most perfect *physique*, drank an unusual quantity of iced water one hot summer's day, and died three days afterwards from paralysis of the stomach.

Dio Lewis, in his work on "Our Digestion," writes as follows:

"Dr. Beaumont makes an interesting statement illustrating the influence of cold drinks upon the digestion. He placed his thermometer in St. Martin's stomach,* and found the temperature 99°. A gill of water at the temperature of 55° was introduced. As soon as it was diffused over the interior surface, the temperature was reduced to 70°, at which it stood a few minutes, and then began very slowly to rise. It was not until thirty minutes had elapsed, and all the water had been for some time absorbed, that the mercury regained its former level of 99°.

"When we reflect that in this case there was but a single gill of water and the temperature was 55°, which hardly deserves the name of cold, we shall not hesitate in pronouncing upon the habit of drinking the usual quantities of iced water with our meals, or that of consuming, at the end of a full meal, a dish of ice-cream. When we remember that a temperature of 99° is absolutely required to carry forward the process of digestion, can we doubt, if a gill of water at the temperature of 55° produced such an effect upon St. Martin, a person of rarest vigor of health, what would be the influence of a

---

* The author would explain, in reference to this apparently rash remark, that Alexis St. Martin was wounded in such a manner by a gunshot that the stomach was exposed; the edges of the wound cicatrized, leaving a permanent fistulous opening leading into the cavity of the stomach. The orifice, usually closed, could be opened, and the process of digestion inspected. Dr. Beaumont made a long series of experiments with St. Martin, most valuable and interesting to the medical profession.

pint of iced water on the stomach of a person of weak digestion."

A more senseless custom still is to drink iced tea. The icing of tea serves to precipitate the tannin, and this is taken into the stomach as an insoluble substance.

SPIRITUOUS LIQUORS.

As medicinal agents I have nothing to say against alcoholic drinks. Strychnine, arsenic, opium, belladonna, and perhaps even calomel, as medicines, may be serviceable also, when prescribed in proper quantities by competent physicians.

Regarding the action and effects of alcohol when taken habitually, I have quoted from Professor Youmans* and others. (See Appendix, page 205.)

I will only introduce here the theory of an able physician who has for many years been at the head of an inebriate asylum in Connecticut. He says that in producing drunkards a fault equal to and possibly greater than the dram-shop influence exists, and that is the custom of habitually serving highly seasoned food at the home table. This creates the appetite for stronger stimulants, which grows and becomes morbid by continued indulgence. The dumb animals, he says, live on simple diet, without condiments, and continue to take the same simple food throughout life that was taken in youth. A pickled carrot, a peppered clover, spiced cornmeal, or a tobacco leaf would be spurned by an intelligent-eyed ox, as an insult to his natural understanding. The sentiments of any other animal (except the human species) would be the same on the subject. Result—scarcely any stomachic disorders among animals.

---

* "Household Science."

The mouth or taste is the heaven-given sentinel to guard that royal domain, the stomach. Give to a young child a condiment and the sentinel rejects it, until by continued solicitation, and by habit, the taste becomes perverted. The stomach gradually acquires an unnatural and unhealthy desire, the same desire that in a strong degree can only be satisfied with fiery liquors. Pepper, Dr. Foote considers the most pernicious of the spices—perhaps because it is more generally used. Tea and coffee are also rejected from his dietetical *repertoire;* in fact, he would discard all articles which tend to excite irritation or create unnatural stimulus.

For the object of mere gustatory pleasure, the doctor insists that the man with healthy appetite enjoys more a simple crust of bread than the epicure with inflamed and calloused stomach can his fiery *ragout.*

In other words, what begins with spice, pickles, etc., ends with whiskey. The man with a healthy stomach will no more crave whiskey than the ox will crave tobacco. In the treatment of inebriates, besides the general remedies administered for febrile conditions, his chief point is to so regulate the diet that the stomach will gradually become accustomed again to simple food, in the same manner that it became accustomed to the stimulating food. When this point is reached the patient is considered cured.

MALT EXTRACT.

Several preparations of malt extract are offered which are valuable in an alimentary point of view, for aiding in the digestion of starch or farinaceous foods.

Malt is made by allowing barley to germinate, and the germination is arrested at a certain temperature. As a result a peculiar nitrogenous principle called diastase is developed, which has the power of converting starch into dextrine and sugar. An infusion of malted barley is reduced to a syrupy consistency, by a low temperature, without impairing the fermenting power of the diastase, and this is called malt extract.

When the digestive powers are weak the extract is often valuable, although it should be taken with farinaceous food, or just after.

The malt extract is also indicated when the mouth is dry, denoting feeble action of the salivary glands. Dr. Roberts suggests that the extract should be spread upon bread and butter, or used to sweeten puddings and gruels.

MILK.

The value of milk as a food cannot be exaggerated. It is a complete diet in itself, containing in proper proportion everything necessary for sustaining life.

In a sanitary point of view the world would be better off if a larger proportion of milk were taken for daily food, and the amount of animal food and of tea and coffee were correspondingly reduced. Milk is not only nourishing, but stimulating; and the natural stimulus resulting from assimilable food is the only healthful and desirable one.

Many diseases, such as rheumatism, dyspepsia, gastralgia, chronic diarrhœa, consumption, etc., are relieved or cured by a diet composed partly or entirely of milk. The milk treatment, as practised in different parts of Europe, has been very successful.

In perfect health, good pure milk is almost always digestible. There are a few with whom it disagrees.

The addition of lime-water will correct it for persons inclined to acidity of the stomach. Skimmed milk will be more beneficial to those who require less fat. When milk is found to be indigestible the difficulty is generally obviated by taking it mixed with starch or grain foods—for instance, with rice, porridge, bread, etc.; or it can be boiled and thickened with a little barley flour, etc. The reason is explained by Dr. Eustace Smith, page 216.

It is preferable to give milk to diabetics in the form of koumiss, which contains no sugar.

In typhoid fever it should be either administered peptonized or in the form of fresh koumiss. This prevents the formation of curd, which is irritating to the bowels in that disease.

Milk in its acid state and buttermilk are nourishing and beneficial in febrile conditions.

Cows' milk is not always of uniform quality. That of the Alderney cow yields the largest proportion of butter. The feeding, too, influences the quality of milk; for instance, with dry food, the milk is relatively richer in solids, and with good grass it abounds in fat.

Water constitutes nine tenths of milk; the remainder consists of albuminoid or the muscle-building principle, caseine (the curd which is used in making cheese), the carbonates or heat-producing principle (the butter and sugar). Then there is some mineral matter—the phosphates. The sugar is called lactine, and by fermentation or souring it is converted into lactic acid.

When the "milk cure" is resorted to, the patient should gradually leave off his ordinary mixed diet until he reaches an exclusively milk diet.

Mitchell formulates his method of administering a milk diet as follows:

"My own rule, founded on considerable experience, is this: Dating from the time when the patient begins to take milk alone, I wish three weeks to elapse before anything be used save milk. After the first week of the period I direct that the milk be taken in just as large amount as the person desires, but not allowing it to fall below a limit which, for me, is determined in each case by his ceasing to lose weight. Twenty-one days of absolute milk diet having passed, with such exception as I shall presently mention, I now give a thin slice of stale bread, thrice a day. After another week I allow rice once a day, about two tablespoonfuls, or a little arrowroot, or both. At the fifth week I give a chop once a day; and after the sixth week I expect the patient to return gradually to a diet which should consist largely of milk for some months."

Dr. Barthelow's rule is gradually to add other diet, after the cessation of symptoms for which the milk treatment was instituted.

Milk and animal food, or milk and acid food, should not be taken together. Persons desiring to take a partial milk diet can take milk and farinaceous food for breakfast and for lunch or tea, and omit it at dinner, which may be a meal of meat and vegetables.

Milk should be taken by the invalid slightly warm. No doubt the natural warmth of the milk when fresh from the cow is the best.

Dr. Dobell, in his work, "Diet and Regimen," says: "Now the nearest approach to a pancreatic emulsion is what may be called nascent milk, by which I mean milk just secreted—milk that flows from the mammary glands as it is formed. . . . In this the emulsification is finest and most perfect, but every minute that elapses after the milk is secreted deteriorates this perfection of emulsification, until, as we know, when allowed to cool, the cream separates from the water of the milk, etc."

Milk can be kept fresh for a long time if placed in well-scalded and perfectly clean glass jars, which can be hermetically sealed by drawing patent wire clasps over the glass tops.

On a journey to Europe some acquaintances took

milk and cream in glass jars, in the way described. The last day of the ocean voyage it seemed as fresh as when leaving New York. It was, of course, kept in the ice-closet.

Glass jars and bottles are now in general use at the best dairies in New York.

BUTTERMILK.

Buttermilk contains, the same as skimmed milk, the full nourishment of the milk without the fat; however, it retains a very small proportion of fat, less than skimmed milk. It is very beneficial in some weak conditions of the stomach (dyspepsia, fever, etc.).

Dr. Ballot, of Rotterdam, has had much to say about the value of buttermilk in the treatment of infants for summer complaint, cholera infantum, etc. Koumiss, or peptonized milk, might be found equally efficacious, and possibly preferable, in many cases.

WHEY.

Whey is almost without nutritive value. As a drink in febrile or inflammatory conditions it is refreshing and often beneficial. It is sometimes recommended to persons who find difficulty in retaining food in the stomach. However, in such cases, koumiss would probably be of greater value.

ANIMAL FOODS.

Of all the animal foods, beef is the most important. It is very digestible, and because of its fine texture and richness in red-blood juices, it furnishes more nutriment in proportion to weight than any other meat. Like bread, it never palls on the appetite. The quality of beef depends much on the age and manner of feeding the ox. To be at perfection the animal should be four years old, not worked, and partly corn-fed.

Mutton is generally more digestible than beef, it and venison being regarded as the most digestible of all the meats. It is popularly supposed to be a lighter food than beef, the latter being adapted to physical exercise, while mutton is rather a food for persons of sedentary habits, and for invalids. Dr. Smith, in an interesting work on "Foods," says that Kean suited the kind of meat which he ate to the part which he was about to play, and selected mutton for lovers, beef for murderers, and pork for tyrants.

Mutton broth has less nutritive value than beef broth.

Venison. When sufficiently hung and tender, venison outranks all meats in point of digestibility. It is also palatable and highly nutritious.

Veal and Lamb. Although the flesh of young animals is more tender than that of old, it is less digestible and less nutritious. The tissues of young animals are more gelatinous than those of the adult, the latter containing more of fibrine and of the flavoring principle, omazone.

Pork. Unless it be a small, thin slice of breakfast-bacon taken in the way of an appetizer, pork should be excluded altogether from the invalid's dietary. Although it is a convenient and inexpensive meat and an appetizing one for many, and perhaps an unobjectionable one for laboring men, yet, on account of the uncertain feeding of the animal, and the hardness of its muscular fibre, it is doubtful whether pork should be used at all by people of sedentary habits.

Lard and pork have seemed indispensable for frying purposes, and for larding and seasoning. It must be taken on trust, however. In the Southern States many are using cotton-seed oil, which has quite the flavor of olive oil, for cooking—using it in place of lard for everything for which lard is used. It is now sold in all the large cities, and in time will undoubtedly be very generally used.

The table inserted below, giving the relative nutritive and other values of the five animal foods principally used, is taken from Dr. Bellow's "Philosophy of Eating."

In one hundred parts are—

	Mineral matter, or food for the brain, etc.	Fibrin and albumen, or food for muscles and tissues.	Fat, or food for heat.	Water.
Veal	4.5	16.5	16.5	62.5
Beef	5.0	15.0	30.0	50.0
Mutton	3.5	12.5	40.0	44.0
Lamb	3.5	12.0	34.0	50.5
Pork	1.5	10.0	50.0	38.5

Undoubtedly too much meat is generally eaten by persons of sedentary habits, resulting in dyspepsia, gout, etc. In cold weather, and with much physical exercise, it can be freely taken, but in temperate or warm weather a greater proportion of cereal food would improve the general health. An analysis of the two kinds of food shows a similar composition. The muscle-making elements in beef, the fibrine and albumen, correspond with the muscle-making elements in wheat—gluten and albumen; and they so agree in chemical composition as to be considered mere modifications of the same substance, and, being dried, contain principally the same elements in the same proportions.

Thus the popular idea that grain food is not so strong and nourishing as animal food is erroneous. The meat of animals is nitrogenous and therefore muscle-building. So are the grains nitrogenous and muscle-building. Dr. Radcliffe, in an interesting article in the *Popular Science Monthly*, says: "It is impossible to distinguish between the albuminose or peptone into which fibrine is resolved in the process of digestion and the albuminose or peptone into which albumen or caseine or gluten or legumine is resolved in this process. It is apparently of little or no moment whether these various nitrogenous articles of food are derived from the world of animal life or from the world of vegetable life.

"You must allow that an herbivorous animal is not less vigorous than a carnivorous animal; and certainly you would find it difficult to show that man, who can live and thrive under the most dissimilar circumstances, upon almost any kind of food, is vigorous in proportion to the amount of meat he contrives to consume."

Meats should not be served to invalids cooked a second time. The flavoring principle, osmazone, is dissipated after the first cooking, and the meat must depend upon outside seasonings for flavor. The tissues are also less tender.

Salted Meats.

On account of the toughness of fibre resulting from the curing process, these meats are difficult of digestion, and should never be used in the sick-room.

Fish.

Fish is a nourishing and digestible food for convalescents if served quite fresh and broiled or boiled. It affords a pleasant change of food about once a week.

Oysters.

Oysters are nutritious and generally well-borne by delicate stomachs. Dr. William Roberts, in his work "On the Digestive Ferments," advances an interesting theory in relation to oysters as a food. He claims that the effect of cooking is to diminish their digestibility, which would make oysters the exception in this respect among the articles that furnish albuminoid matter. He explains his reasons by saying that the fawn-colored part of the oyster, containing about half its substance, is its liver, composed partly of glycogen. Associated with this, but withheld from actual contact with it during life is its appropriate digestive ferment—diastase. Mastication mixes these constituents and they are digested without other aid. Cooking destroys the digesting properties of the diastase, and then the oyster has to be digested like other food—by the eater's own digestive power.

Other authorities question and doubt Dr. Roberts' theory. The excessive use of condiments—salt, pepper, lemon-juice, and vinegar—more especially pepper, combined with imperfect mastication, may possibly impair the wholesomeness of raw oysters to many persons.* The hard portion, or muscle, which fastens the oyster to the shell should be removed in all cases when served for weak stomachs.

The author would recommend oyster soup, properly prepared (the oysters slightly cooked), as the best mode of administering oysters to an invalid. The flavor of the juice and the extra nourishment furnished by the cream or milk used, together with the advantage which foods served warm afford to digestion, would be good reasons for preferring oyster stews or soups.

Dr. Bellows, in "The Philosophy of Eating," speaking of oysters, says: "They have not, as food, the muscle-making elements of the crustacea or other active fish; and although their chemical composition indicates phosphatic salts, they are mostly salts of lime, which go to form the shell and to make bone rather than a food for the brain and nervous system. Oysters, therefore, are very unsatisfactory food for laboring men, but will do for the sedentary and for a supper to sleep on. They contain but $7\frac{1}{2}$ per cent. of solid matter, including fibrin, albumen, gelatine, mucus, and osmazone; and

---

* Whatever the cause, the fact remains that many persons find raw oysters quite indigestible.

much of that is gelatine, which affords no nourishment, while butcher's meat contains on an average 25 per cent., and the poorest fishes contain 14 per cent., of pure nitrates. The nitrates in oysters are in the form of albumen, like the white of an egg; they are, therefore, more easily digested in a raw state than when cooked, but when stewed are not indigestible."

FAT.

We have heretofore too little appreciated the importance of fat in our dietaries. Without knowing why, fat has generally been considered unwholesome, tending to produce biliousness, corpulence, and heat, besides being a general clog and burden in all digestive processes. Oil has been avoided; butter on bread has been scraped down to the smallest quantity, and the fat of meat has been sedulously trimmed.

Fat is as necessary to the system as the muscle-making properties of foods. It not only serves to produce heat, but has an essential share in the tissue-making process. It does not produce the material, but influences the assimilation of the other principles of food by well-established processes. Although it is essential to the formation of every structure in the body, it is an especially essential constituent of the brain and nervous system. A diet with a deficiency of fat tends to produce diseased conditions in the direction of scrofula and consumption. Cod-liver oil is not properly a medicine; it is a fatty diet given with a view of supplying what is supposed to be lacking in the system. It is affirmed that if one takes and assimilates a sufficient quantity of fat in the ordinary diet, one is not liable to have consumption or nervous diseases.

In foods supplying all the necessary elements for sustaining life, fat constitutes a considerable proportion

—for instance, milk, eggs, etc. The yolk of the egg is about one third fat.

Dr. Radcliffe says, "There is no essential difference as to the chemical composition between vegetable albumen and fibrine, and legumine and oily matters, and animal albumen and fibrine, and caseine and oily matters; there is no perceptible difference in the albuminose or peptone into which the vegetable and animal nitrogenous substances are alike transformed in the process of digestion; there is no difference in the way in which the vegetable and animal oily matters are emulsified and then taken up directly into the general circulation of the blood."

Another writer says: "If the inhabitants of the Arctic regions gorge themselves with animal fat, those of warm countries take the same thing in vegetable oils. In most warm climates olive-oil is taken, and in India ghee, with no inconvenience to digestion and with unmistakable benefit."

An interesting article on the subject of fats, by Dr. Radcliffe, was published in the *Popular Science Monthly* (March, 1883). It is in the form of a dialogue between a physician and a young man who had eaten a breakfast of lean meat and toast in anticipation of a hard day's rowing. The physician explains to the young man his mistake, and shows that, as force-producing agents, fat and oil are as necessary as fibrine or albumen.

He also says: "I find that very many persons suffering from various chronic disorders of the nervous system have abstained from the fatty and oily articles of food, and that their state is almost invariably very much changed for the better when induced to take what they have avoided."

Because we have, perhaps, been mistaken in taking too little fat in the past, it is not recommended that too large a quantity be taken in the future.

Pavy says the supply ought not to be less, even with inactivity, than one ounce daily, and that about two and a half ounces will constitute the average amount in the dietaries recommended for working people.

Fresh milk furnishes fat in proper proportions. Cream and butter furnish the most assimilable fat. Bread generously buttered (not too much so, however), meat with streaks of fat, and the oil dressing on salads will ordinarily afford a sufficient supply. Pork fat is the most objectionable of the fats to persons of sedentary habits.

Dobell says: "When it is necessary, for any special object,* to reduce the quantity of carbon taken in the aliments, this can more safely be done by diminishing the saccharine, amylaceous (sugar and starch) matters, than the fat."

EGGS.

Eggs contain all that is required for the building and maintenance of the body. They are, therefore, a most invaluable article of food. The white is almost pure albumen and water, and the yolk contains the fat and other necessary constituents. They are more easily digested when taken raw or slightly cooked, as described for poached eggs (cooked in water below the boiling-point). Continued boiling, or cooking in any manner, toughens the albumen and renders it difficult of digestion. Indeed, a valuable cement is made by thickening the white of egg with powdered quicklime, and heating it. The whole egg can be made hard and tough enough by heating to become a cement of itself.

RICE.

Rice is very rich in starch, and poor in fat and albuminous matter. It contains less than half the muscle-

---

* For instance, to reduce corpulency.—ED.

supporting elements of wheat, and only one fourth as much of those going to support the brain and nerves. Rice-eaters are, therefore, feeble and indolent. The deficiencies, however, can be supplied by cooking it with milk or eggs.

It is very digestible, requiring only little more than an hour for the process. In weak conditions of the stomach and bowels it is valuable. Rice-water, a thin mucilage, is a drink often administered with benefit in fevers and in inflammation of the bowels.

CORN-STARCH AND ARROW-ROOT.

Corn-starch, and arrow-root, composed chiefly of starch, are inadequate to sustain life without the addition of milk or other nutritive substances.

SAGO AND TAPIOCA.

These are also starch foods, and they rank very low in an alimentary point of view. They are chiefly used as pleasant additions to custard puddings, and as a thickening for soups.

BEANS AND PEASE.

These are rich in nutritious material. Their muscle-making element is not gluten, as in the grains, but caseine, as in cheese, a substance not so easily digested as gluten, and therefore only adapted to strong and active persons, with good powers of digestion.

GELATINE.

Jellies and blanc-mange made with gelatine are very appetizing, but cannot be relied on as furnishing nourishment. Calves'-foot jelly was once regarded as a valuable dish in the sick-room. It is a very pleasant vehicle for serving wine or milk; but, beyond this, it is now believed to be valueless by all the best authorities. Several years ago a committee was appointed by the French Academy of Sciences to ascertain the dietetical value of gelatine. This was on account of the fact that gelatinous extract of bones was being fed to the inmates of hospitals with apparently deleterious results. The commission, with Magentie at its head, reported gelatine to be substantially worthless as a diet.

TOMATOES.

The tomato, according to Dio Lewis, is a medicinal vegetable containing some amount of calomel—enough to produce a degree of salivation if used too freely. He thinks the tomato should be used moderately in cooked form, as a sauce, etc. He has known, in his practice, of patients suffering with sore mouths, tender and bleeding gums, with loose teeth, and with piles, produced by the immoderate use of tomatoes.

However wholesome a certain amount of cooked fresh tomatoes may be, the physicians generally denounce the use of them when put up in tin cans. The tendency of the acid of the vegetable is to corrode the tin, and thereby, to some degree, poison the tomatoes.

FRUITS—GRAPES, BANANAS, ETC.

Fruits are cooling, aperient, and nutritious, and are almost as necessary to a healthful dietary as the grains, especially in warm climates. They cool and refresh us in summer, supplying grateful acids and fluids. They are grown on all inhabitable parts of the earth, and many of them can be kept in all seasons. Different varieties of fruits follow each other in close succession during the season of growth, the acid fruits coming generally in the spring, when the system needs anti-bilious food, after the winter dietary.

Next to the apple, the king of fruits, the grape is probably the most valuable in our climate. Its beneficial action seems almost medicinal. After eating the grape regularly for some time, when it is fresh picked from the vine and redolent of the sun, general exhilaration is produced; the blood seems richer and a healthy glow of color comes to the cheeks. Besides the tonic effect, the grape contains much nourishment.

They have in France, Switzerland, and Germany what are called grape-cures, where persons suffering from dyspepsia, scrofula, gout, and cutaneous diseases are treated during the grape season with much success. Patients eat the grapes to repletion several times a day, and at regular intervals, generally taking nothing with them but bread-and-butter and water. Dr. Barthelow says, however: "The influence of change of air, of scenery, and of the hygienic rules enforced at these resorts should not be ignored in an estimate of the value of the method." Hot-house grapes, and the California grapes after transportation to the Eastern States, will not answer the purpose, nor take the place of the Isabella, Concord, Catawba, and other varieties grown in the open air, fully ripe and fresh from the vine.

Another nutritive fruit is the banana. It contains a large percentage of starch and sugar, and enough nitrogenous matter to make it of alimentary value. It is similar in composition to the potato. In some tropical countries it is much used as a food. On a plantation in Cuba the owner took us to see the negroes prepare their dinner. A huge iron pot, hanging over burning fagots, was filled with a combination of materials making a sort of *ragout*. The chief ingredient, they told us, was the banana.

For invalids, berries with hard seeds—strawberries, raspberries, etc.—are often indigestible. Many of our marketable strawberries are so very acid and devoid of flavor, that they, especially, cannot be recommended to invalids.

Stewed fruits (compotes) are very wholesome and beneficial for almost any one. They should be served in some form every day, provided a laxative diet is not at the time objectionable.

When oranges, and they are especially excellent in all febrile conditions, are administered to invalids, they should be quite sweet. There seem to be as many varieties of oranges as of apples. Although a juicy, crisp, moderately sweet, and well-ripened apple is the most wholesome and digestible of fruits, there are apples which can defy the ordinary stomach, and which set the teeth on edge to even think about; so it is with some oranges, which are only fit for orangeade. The sweet, juicy, thin-skinned, little Florida orange, and the more rugged skinned, though juicy and sweet, Havana orange can be judiciously given to almost any invalid, while their more common and acrid relatives should be as carefully avoided.

Baked apples served with cream and sugar are a standard dish for the sick-room. They are digestible, laxative, and very wholesome.

The dried fruits, especially the California dried pears and the white apple-chips, are very refreshing and safe, and should be more used when fresh fruits cannot be obtained.

If fruits are not quite ripe, or do not agree with one, cooking them with sugar increases their digestibility.

Acid fruits put up in tin cans are exceedingly doubtful. If they taste of the tin, they are not at all doubtful. Avoid them. Probably, in the future, tomatoes and acid fruits will be generally put up in glass jars, if something else less breakable than glass, and without

Sea-Moss Farine and Sea Moss.

An article was sold several years ago, at all the grocers, called sea-moss farine. It was a most excellent preparation, especially valuable for invalids, and could be made into various blanc-manges and puddings, according to directions accompanying the packages. I have tried in vain to obtain it within the last two or three years, and it seems to be out of market. I hope very much to see it in market again, as it is an especially valuable health-food.

Sea moss is very nutritious, exceedingly digestible and wholesome, and can be used to advantage for almost any invalid. Its flavor takes one to the sea-shore, it matters not how far away. The blanc-manges made from the Irish and Iceland mosses are especially good.

THE NEW HEALTH-FOODS AND OTHER GRAIN PREPARATIONS.

The new methods of preparing cereals by the Health-food Company of New York have produced the most gratifying results. These foods are of inestimable value to the invalid. Indeed, they constitute a pleasant and wholesome diet for any one. Their use tends to preserve health, and preservation is far pleasanter than restoration.

The manufacture of foods after methods based on careful scientific investigation, specially adapted to the needs of different individuals and diseases—for instance, foods for the corpulent, or the excessively lean, for infants, for diabetics and dyspeptics, and for persons generally debilitated, where serviceable treatment must be chiefly dietetic, is of especial value.

Heretofore in the treatment of diabetes, where the patient is obliged to eschew all foods containing starch or sugar, thereby depriving him of bread and all grain preparations, the physician has had much embarrassment. The "Diabetic Food," consisting of gluten, which is nutritious and very digestible, is a boon to these sufferers.

It is known that heretofore in milling wheat the most nutritive portion of the grain, the gluten, lying next to the hull, was removed. The white flour, making bread quite beautiful in appearance, is chiefly composed of starch, and is incapable of sustaining life.

A distinguished physician said, "The intelligent

farmer knows how to feed his land, his horses, his cattle, and his pigs; but not how to feed his children. The fine flour, containing neither food for brain nor muscle, he gives to his children, and the whole grain or the bran and coarser part, containing food for brain and muscle, he gives to his pigs."

Formerly, in the preparation of Graham flour and cracked wheat, although the full nutriment of the grain was preserved, the hull, a woody, fibrous skin, was retained. This proved to be irritating to some delicate stomachs, although authorities say that it serves a good purpose for vigorous persons, viz., of promoting by a healthy irritation the secretions and motion of the bowels.

The Health-food Company manufacture, besides flour with its full richness of gluten, coarser preparations of the cereals, such as granulated wheat, oats, barley, corn, etc., with the silicious skin removed.

Some of the articles prepared by this company, which may be commended as deserving, are:

The COLD-BLAST WHOLE WHEAT FLOUR—represented to contain the full nutrition of the grain.

PEARLED WHEAT.—The whole grains of best wheat denuded of their bran coats.

GRANULATED WHEAT (Coarse)—which takes the place of the ordinary wheaten grits or cracked wheat, is also prepared without the bran coats. The last two preparations make an especially palatable dish, prepared according to the Vienna Bakery receipt given on page 128.

GRANULATED WHEAT (Fine)—takes the place of Graham flour. This flour can be employed in the various ways in which Graham flour is used, viz., for making bread, crackers, mush, pancakes, croquettes, puddings, thickenings for soups, sauces, etc.

This company prepares also WHITE WHEAT GLUTEN, a concentrated, digestible, and nutritious food. Being free from starch, it is recommended to those suffering from dyspepsia, diabetes, and Bright's disease, and also as an anti-fat diet.

It was not found to be very palatable by the author. It can be made into bread "gems," mush, pancakes, puddings, etc. If some starch be unobjectionable, the gluten is much pleasanter to the taste when mixed with flour, rice, or barley.

GRANULATED BARLEY.—Bellows says of barley: "This cereal compares well with wheat in nutritive elements, but does not form bread; is used for making barley-cakes, which are valuable for persons inclined to constipation, containing, as it does, more of waste which is the natural stimulant of the bowels. Barley is peculiar also for its richness in phosphates, having more than twice the amount contained in wheat; and therefore might be made useful to literary men of sedative habits, adapted as it is both to promote the action of the brain and bowels."

The pearl barley ordinarily used in soups is a grain that does not dissolve in the cooking process, and is quite unfit to be used in the sick-room.

The "Granulated Barley" of the Health-food Company dissolves as easily as rice. It is a valuable preparation and can be used to great advantage in a variety of puddings, the best being a *soufflé* pudding (page 192); also in pancakes, gruel, thickening for soups, blanc-mange, etc.

The best preparation I have seen for making barley gruel (one of the most valuable of gruels) is Robinson's barley flour. It is manufactured in England, but is very generally sold here by the druggists. It is exceedingly palatable and valuable for invalids.

PEARLED OATS.—Good for making porridge; also the GRANULATED OATS, admirable for puddings, gruels, etc., and the OAT FLOUR, especially good for gruels. These are all articles which can be used beneficially in many different ways.

"CEREAL COFFEE"—made of barley and wheat gluten parched. It is a good substitute for tea and coffee. It has some of the coffee flavor and is without more stimulant than is imparted by any nourishing drink.

Among other of the health-food preparations are crackers made of the cold-blast flour, gluten, oats, granulated wheat, etc.

The manufacturers of what is known as the new patent-process flour claim that it also contains the full gluten of the grain. The flour is necessarily of a creamy color, gluten being light brown in appearance. This flour can be obtained of grocers in all of the large cities. If the flour sold for the "new-process" flour is purely white, it is not genuine. If the necessary amount of gluten is retained it must color it to some extent, indeed, to the extent of giving it a decidedly creamy hue.

There are agencies in the large cities for cereal foods (oatmeal, barley, groats, hominy, cornmeal, etc.) prepared at Akron, Ohio, which are most excellent. The Graham flour from this source is especially fine.

The best oatmeal which can be obtained is the imported Irish oatmeal. It can be purchased of the first-class grocers in New York, but the author has not been able to find it elsewhere. It is more palatable than the Scotch or American oatmeal, the grain being much larger.

The concentrated foods so industriously advertised are not recommended by the authorities. A certain amount of bulk is necessary, and the less nutritive portions of food perform a very necessary function in the process of digestion.

KOUMISS.

This nutritious beverage, made of fermented milk, has been hitherto comparatively unknown in our country. It has been used for centuries in Tartary and in Asiatic Russia. It is there chiefly made of mares' milk (see Appendix). Mares' milk differs from cows' milk, the former possessing (according to Pavy) a smaller amount of nitrogenous matter and butter, and a much larger amount of sugar. By adding sugar to cows' milk a koumiss may be obtained superior in its nutritive properties to that made of mares' milk.

Koumiss is of incalculable value for almost all invalids, containing the full nutriment of milk and the stimulating qualities of wines and liquors without any ill-effects.

Dr. Dobell, of London, in his valuable work on "Diet and Regimen," says: "Koumiss, when properly prepared, is a highly refreshing, effervescent preparation of milk obtained by a natural process of fermentation, in which the albumen and casein are partly digested, while its abundance of free carbonic acid makes it sedative to the most irritable stomach, so that it has succeeded in numerous cases, recorded by medical practitioners, where stimulants, beef-tea, and rectal enemata, aided by the most varied pharmacopœial treatment, had alike failed.

"Its chief qualities are:

"(*a*.) Its agreeable, refreshing, and highly digestible character.

"(*b*.) Its attested and rare powers of nutrition, in the most desperate cases of emaciation, chronic vomiting, dyspepsia, gastric pain, and irritability, and of debility following acute or accompanying chronic diseases.

"(*c*.) The avidity and pleasure with which it is drank by children, women, and men, in health and disease, and in its remarkable success in allaying vomiting and gastralgia, and in restoring the nutrition."

Dr. Roberts Barthelow, in his "Materia Medica," says: "Koumiss differs from whey in containing the nutritive constituents of milk, and from milk itself in the important respect that it is, in addition, an effervescing, alcoholic fluid. . . . The tolerance of the stomach to koumiss is remarkable, even in cases of gastralgia. It improves the appetite, and excites the action of the kidneys. The patients experience a pleasing exhilaration, due probably to the combined action of the carbonic acid and the alcohol. It also causes somnolence during the day, and favors sleep at night, without leaving any after headache. Its most important action is the increase of the body nutrition. . . . Koumiss possesses great value in the treatment of consumption, chronic bronchitis, the low stages of fever, the stage of convalescence from acute diseases—in fact, in all adynamic states in which the combined effect of alcohol and nutrients may be desirable."

Jaqielsky says that he has had patients gain as much as ten pounds a month, when no other food was taken.

Koumiss, in its administration, may be given like milk or beer. In extreme cases of feebleness of digestion, this being the only food, a glassful every two hours would be sufficient. With increased facility of digestion and assimilation from a quart to a gallon a day may be taken. When served with other food, a glassful can be drank before or after a meal as preferred. It is a food in itself—a solid food, like milk, containing all the elements or requisites of nutrition. The cascine of milk turns into curd in the stomach, and leaves a solid residue. It is estimated that each quart of koumiss contains four ounces of solid food.

After such a *richesse* of authority (and there is much more before me) it would seem unnecessary to mention that I have become enthusiastic as to the merits of koumiss, after having seen its almost miraculous effects upon a member of our own family. In this case no food whatever seemed assimilable until koumiss was prescribed. This led me to investigate and experiment with the making of koumiss with results which I hope will prove as satisfactory to others as myself.

There are two kinds of koumiss—one quite acid, like that generally sold at pharmacies in the large cities; the imported koumiss is also quite acid. The venders of this koumiss say that it improves with age, that two or three years old being considered especially good. This acid koumiss would be indicated in cases of fever, rheumatism, etc., when acid drinks, such as buttermilk, lemonade, etc., are relished and required.

For a more ordinary and general drink the sweet koumiss (perhaps it can hardly be called sweet, as the flavor is pungent, not unlike beer), made as imperfectly indicated in many of the medical works is preferable. This is at its best from four days to a month old. In my own experience, there were several days when our invalid craved something acid. Not having the proper acid koumiss at hand, it was found that some koumiss which had curdled and soured (this comes from the bottle as effervescent as that which has not curdled), agreed with her perfectly when buttermilk proved indigestible. After two or three days the appetite no longer called for acid, and the sweet koumiss was more assimilable as well as better relished.*

When it is desired to give koumiss to babies, they can either suck it from the end of the champagne-tap, the screw being turned very slightly, or a little koumiss can be drawn into a pitcher and poured from one pitcher to another until most of the gas has escaped. The infant can then drink it as milk.

To Make Koumiss.

The making of koumiss is very simple. It requires perfectly fresh milk, good yeast, a little sugar, strong bottles (those used for champagne, beer, etc.), a corking-machine (price, fifty cents), a little tuition in the professional manner of tying corks in bottles, a thermometer, a funnel, a cold, dark place in a cellar answering the purpose of a beer cave, and *voila tout*—not quite all though, for if one's life or the roof of the house is regarded of value, a bottle of koumiss should not be opened without a champagne-tap.

Fill a quart bottle about three quarters full of fresh milk, and add a tablespoonful of fresh (brewers') lager-beer yeast, and a tablespoonful of sugar-syrup (the syrup is made allowing three lumps of sugar—little squares of loaf sugar—or a tablespoonful of ordinary white sugar, for each quart of milk; enough water to cover the sugar is added, and it is boiled a couple of minutes to make the syrup, not allowing it to candy); shake the bottle well for a full minute, to thoroughly mix all the ingredients, then fill it to within two or three inches of the top; shake again, to get all well-mixed. Cork it with a

---

* The author, since writing this, has had occasion to know of several other invalids who have tried koumiss. The very acid koumiss usually sold by druggists was quite unsatisfactory, excepting for temporary use for fever, whereas the fresh koumiss was marvellously successful.

cork a third of a size larger than the mouth of the bottle. The corks must have been previously soaked for two or three hours, *immersed* in hot water over a warm stove, when they become soft; they are then pushed through the corking-machine (see cut) with a hammer, or, better, a wooden mallet; quite heavy and vigorous blows of the mallet on the handle of the machine will not break the bottle, as one might suppose. The corks are then tied. When this operation is all completed, put the bottles in a standing position in an even (or as nearly so as possible) temperature of 52° Fahr.,* where they should remain for two and a half days. Some closed closet or cellar in winter or a refrigerator in summer will generally afford this temperature. This slow fermentation is desirable. At the end of the two days to two days and a half, place the bottles *on their sides* and on the stone-floor of the darkest and coolest place in the cellar—or, in default of such place, in a refrigerator. Many consider koumiss at its best when it is five or six days old, but it can be kept indefinitely if kept in a temperature not above 52°. The colder it is kept without freezing the better. The brewers' lager-beer yeast is decidedly the best for making what I call the sweet koumiss, imparting to it a beer flavor. As the

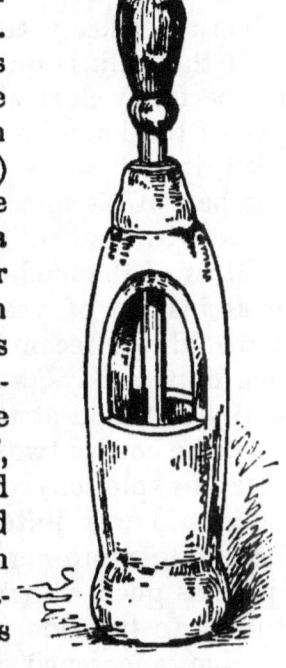

CORK MACHINE.

---

* My first instructions were to leave the koumiss at this stage in a temperature of 62° for the two and a half days, but I have found, by experimenting, that a temperature as low as 52° produces even better results.

konmiss is drawn it should appear in the glass like thick whipped cream. The koumiss will become acid by long standing, or by placing it in a higher temperature.

Very good koumiss can also be made with Fleischman's Compressed Yeast. A fifth of a two-cent cake of this yeast to a quart of milk is the proper proportion. It should be well-dissolved before it is added to the milk. The proportion of sugar or syrup is the same as when the other yeast is used.

If the milk is quite fresh and sweet, and the bottles are perfectly clean and free from acid, there is little danger of the koumiss curdling. If it should become curdled, it can be used for cooking purposes. It makes the best of biscuits, pancakes, or anything which can be made with sour milk.

Most of the medical works advise the use of old koumiss instead of yeast to produce fermentation. This I would not recommend. After the koumiss is made one or two days, a thick curd (the caseine) will generally be found at the top. It is also recommended to turn the bottles two or three times (not shake them, for fear of explosion) so as to mix this curd with the liquid below. I was quite particular about this at first, but, becoming more negligent, found that the koumiss was quite as good without this care. When the bottles are turned to the side (after the two and a half days), the caseine is loosened from the top, and when the koumiss is drawn, the effervescing gas accomplishes the mixing.

To Tie the Bottles.—With a strong hemp twine make a loop as in Fig. 1, page 37.

In Fig. 2, the twine at *a* is drawn up, and in Fig. 3 it is placed over the top of the cork. The two ends, *b, b*, are drawn as firmly as possible under the rim of the bottle, *c*, as in Fig. 3.

The ends, *b, b*, are then tied firmly over the top of the cork, Fig. 4. If the twine is not quite strong, the bottle can be doubly tied.

THE CORKS.—The corks should be obtained at a cork factory or wholesale cork store. The directories in the larger cities will give such addresses. They there cost fifty to sixty cents a gross, instead of a cent

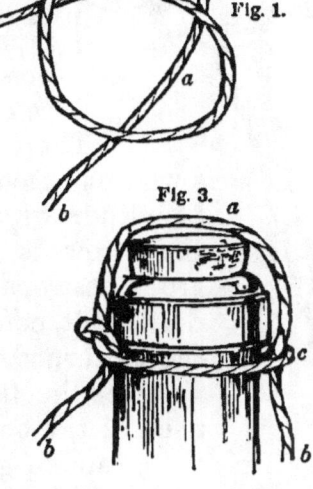

Fig. 1.

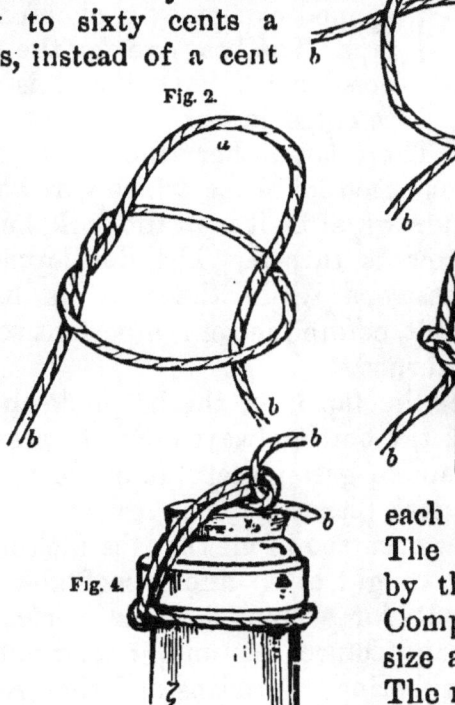

Fig. 2.

Fig. 3.

Fig. 4.

each as at the druggists. The straight cork used by the Anheuser Beer Company is of the proper size and of best quality. The necks of champagne and beer bottles are of the same size, the same cork answering for either.

TO CLEAN THE BOTTLES.—If the koumiss is not acid, merely cleansing the bottles, as soon as emptied, and filling them with cold water will be sufficient. If any acid remain in the bottle, shake it well, half filled with water, with a half teaspoonful of soda added. Pour this out, add another half teaspoonful of soda, fill the bottle with water, and let it remain until it is wanted for use, when it should be rinsed with fresh water.

THE CHAMPAGNE TAP.—It must be repeated that the koumiss bottle should never be opened except by a champagne tap. The best one for the purpose that I have found is represented in cut.

There is another kind of tap with a wire enclosed in a tube, which wire has to be withdrawn after it is in the cork, before the screw is turned. The developments are disastrous while the wire is being drawn out, before the very important screw can be turned.

After the tap is in the bottle, keep the neck of the bottle always down to prevent the escape of gas. Keep the bottle also in a cool, dark place.

CHAMPAGNE TAP.

It has occurred to me that the making of koumiss might often afford profitable employment for women. After perfecting themselves in its manufacture, they might send notices and samples to neighboring physicians and then sell it through the agency of druggists or grocers; the latter having generally better means for the transportation and delivery of articles. The difficulty in procuring quite fresh milk in the large cities might preclude its best manufacture there.

ARTIFICIAL DIGESTION BY MEANS OF PANCREATIC FERMENTS.

IMPORTANT discoveries have lately been made in the matter of supplying artificially digested, or partly digested, food, which is of great benefit in the treatment of certain diseased conditions. The digestive agent is pancreatic juice, or ferment, which can be taken from animals in an active, potent form. This is mixed with milk, milk gruel, milk punch, beef tea, and other foods, as explained in the receipts. Such digested food is especially indicated when there is an inability to digest the caseine of milk, or starch or fats, as often occurs with infants unable to retain milk in the stomach, and with consumptives who cannot digest fats. It is also indicated in cases of extreme emaciation or weakness, in cases of typhoid fevers,[*] and especially in gastric troubles brought on by alcoholic excesses.

It is probably better to resort to artificial digestion only in extreme cases, where exercise and bracing air cannot accomplish their usual results in aid of natural digestion.

Pepsin for stomachic indigestion has long been in use. Much of the digestive process, especially in the case of fats and starches, takes place when the food has left the stomach and entered the large intestine. This

---

[*] The ulcerated bowels, common to typhoid fever, must not be exposed to the irritation of foods that leave a solid residue after digestion. The curdling of the caseine of milk can be prevented by giving it already digested (peptonized).

may be called intestinal digestion. It is here that the pancreatic ferment does its work.

For information on this subject, viz., the practical use of pancreatic extract and its action on the human system treated philosophically, we are indebted to Dr. William Roberts, of Manchester, England. This information was given in a series of lectures before the Royal College of Physicians, which have since been published in book form, entitled "On the Digestive Ferments."

In our own country a preparation of the pancreas, called "Extractum Pancreatis," is made by Fairchild Brothers & Foster, New York City. Mr. Fairchild has published a small book on the subject, having given it a very thorough investigation. His extract is in powdered form, is easily kept, and quite perfect in its results.

Dr. Horace Dobell has also contributed valuable information on the same subject; having, in fact, preceded Dr. Roberts in his publications. His experiments have been chiefly directed to the action of the pancreas on fats.* An article, which can be obtained in most of our large cities (prepared by Savory & Moore, of London), called "Pancreatic Emulsion," *i. e.*, pancreatized suet, cod-liver oil, etc., is the result of his investigations. This aliment is considered especially valuable for consumptives. (See Appendix, p. 212.)

Dr. Dobell says: "Pancreatic emulsion has proved most magical in its effects on miserable, wasted children—children who have been subjected to chronic defects in diet; for instance, when the mother's milk is poor in fat and lactine, or when the child's diet has been de-

---

* The albuminoids and starch have been digested with pepsin and vegetable diastase; no other digestive agent has been found to emulsify fat but pancreatine; the pancreas, however, is the only organ concerned in the digestion of fat.

ficient in milk and fat elements, and the pancreas has been partly paralyzed by prolonged inactivity, causing a kind of wasting (marasmus)."

In the preparation of the various foods with the pancreatic extract, the process of digestion is stopped a little short of completion, to prevent the formation of offensive products which full digestion would develop.

In any of the following receipts the milk or food may be more or less peptonized.* For ordinary cases, especially for infants, it is better to partially peptonize the food. The degree of peptonizing is best determined by the readiness with which the food is assimilated by the patient. To check the action of the digestive ferment, the food, when sufficiently peptonized, can either be placed on ice, which at once arrests all action (and is a commentary on the reckless habit of drinking icewater), or it can be scalded, or brought to the boilingpoint. It is afterwards kept like ordinary milk. Peptonized milk gruel is generally preferred to the peptonized milk.

To Peptonize Milk.

In a clean quart bottle put a powder of five grains of Extractum Pancreatis (about a quarter of a teaspoonful), also fifteen grains of soda† (a pinch), and a gill of water (half a cupful); shake it, then add a pint of *quite fresh* milk.

Place the bottle in a pitcher of hot water, or set the bottle aside in a warm place for an hour, or an hour and a half, to keep the milk warm—about 110°, or the natural temperature of the body. When the contents of the bottle assume a grayish-yellow color, and a slightly bitter taste, then the milk is thoroughly pep-

---

* The word peptonized is used as synonymous with pancreatized.

† A newer preparation of the pancreatic extract comes already mixed with soda.

tonized. When partially peptonized it has no bitter taste, and but little appearance of change. When the milk is peptonized (sufficiently for the patient), either scald or bring it to the boiling-point (to prevent further digestion), or place it on ice until used. It can be taken like ordinary milk. (See Appendix, p. 213.)

Peptonized milk may be sweetened to taste, or used for making punch, with rum, etc., or it can be made into jelly; indeed, it can take the place of ordinary milk in any of the various dishes in which milk is used.

Peptonized Milk Gruel.

Half a pint (a cupful) of well-boiled gruel (of barley flour, Graham flour, or granulated wheat, corn, or oatmeal) is added while still boiling hot, to half a pint of cold milk. The mixture will have a temperature of about 125°; add to this five grains (quarter of a teaspoonful) of the Extractum Pancreatis, and fifteen grains of soda, and let it stand until peptonized, the same as for peptonized milk, both as to making and preserving.

Peptonized Milk Jelly. (Very palatable.)

Ingredients: one pint of peptonized milk heated to boiling; one quarter of a pound of sugar; a half-box of Coxe's or Nelson's gelatine; the juice and the thin yellow cuts of the rind of one lemon; the juice of one orange; three or four tablespoonfuls of Jamaica rum.

Add the sugar and the thin cuts of the rind of the lemon to the milk. Soak the gelatine for half an hour or more, in enough cold water to merely cover it, then add a gill of boiling water, and when quite dissolved add the juices of the lemon and orange, and also the rum. Add this to the sweetened milk when it has partially cooled, and pass it through a little wire milk-strainer or sieve. Pour it into cups or moulds (previously wet with cold water), and set in a cold place.

This jelly can be made of any flavor, with or without wine or spirits. It is very good when flavored with lemon or orange alone, or with lemon or almond extract.

When the milk is thoroughly peptonized (brought to a point when a slight bitter taste is detected), lemon juice or acids will not curdle it, as with the ordinary milk.

The milk gruels can be used as well as the milk itself in making jelly.

GRAPE JUICE.

The value of simple grape juice as a beverage has become but recently known, principal attention heretofore having been directed to its fermentation into wine. For the invalid the simple grape juice is far preferable, the natural tonic of the grape being obtained without the inflammatory effects of alcohol. In flavor the natural bouquet of the grape is preserved. No beverage, aside from water, is more generally wholesome and palatable. In some of the hygienic institutes it is prepared in large quantities and drank in place of tea or coffee at meals.

It was introduced into St. Louis by Dr. Dodds in 1872. I am indebted to her for the mode of its preparation. Its manufacture provides a new industry for the farmers and canning companies, as the use of simple grape juice is destined to become general.

Its preparation is as follows: Take grapes thoroughly ripe and fresh from the vine. The Concord and Isabella are especially good, but any fresh, ripe, and juicy grape may be used. Allow one quart of water to three quarts of grapes freed from the stems. Use no sugar. Let it come slowly to a boil, and when the whole mass is boiling hot strain the juice through a cheese-cloth, flour sack, or other strong cloth. Then return the liquor to the fire, and as soon as it is at the boiling-point again, can it.

The less the fruit or juice is cooked the brighter will

be its color and the better the natural flavor of the grape will be retained. This, like all other articles to be canned, must be at the boiling-point when it is sealed. If the juice is to be used at once it should not be brought to the boiling-point a second time. Use wooden spoons in its preparation, and only glass jars for keeping it. The action of any acid substance on tin is to corrode it and poison the fruit.

Before heating the grapes see that all the necessary preparations are complete, viz., that the jars and covers are clean, the covers fitted, and the hot water ready for holding the jars, etc.

To avoid breaking the jars, manage them as follows: When the grape juice is nearly ready for canning, fill a large wooden tub about three quarters full with water quite hot, but below the boiling-point. Holding the jar sidewise, roll it over quickly in the water, and then set it right side up with the water in and around it. Continue in the same manner with other jars. Place the covers also in hot water. The juice being ready to be canned, roll one of the jars again quickly in the hot water, empty it, place it on a tin platter, and pour it full of the boiling juice, rather slowly at first. Wipe the moisture from the top of the can, adjust the rubber ring, and screw on the top (taken from the hot water and wiped dry) until it clasps the rubber tightly all around. Do it all as quickly as possible. Set this jar aside and proceed in the same way with the others. After the jars are cool enough to handle, screw down the tops again, and when entirely cold give them another twist in order that the sealing may be perfect. The best plan is to let them stand twenty-four hours and tighten them from time to time. Last of all, wipe them clean with a damp cloth, and set them away in a *dark*, cool closet or cellar. If no dark cellar be at hand, wrap the bottles in heavy brown paper to exclude the light. The cooler they are kept without freezing the better.

THE HOT-WATER CURE.

The drinking of simple hot water as a cure for rheumatism, gout, dyspepsia, catarrh, etc., is new and very efficacious. In these diseases there is a sporous condition, or an animal or vegetable growth on the coatings of the stomach or respiratory tubes. The tendency of *hot* water is to produce an irritation and excite an action of the mucous membranes of the tubes and stomach, which throws off or detaches diseased matter. The tendency of *water* is to wash off these impurities and to carry off through the kidneys any effete matter.

The water should be taken as hot as possible. It is often taken in a wooden goblet. It should be taken on an empty stomach, either half an hour before a meal or two hours after. Two or three quarts a day are taken by some, although ordinarily a glassful (a half-pint) is taken half an hour before breakfast, again at 11 o'clock, and again at 4 P.M.

Hot water taken in this manner, as a remedial agent, is a comparatively new discovery. It was found that rheumatism, gout, etc., were cured at the Hot Springs of Arkansas by the patients drinking quantities of the hot water at the springs. Experiments showed that any pure hot water was as good as that from these famous springs, the diuretic effect being what was required.

A physician in New York told me that the hot-water mania had been carried too far. He never prescribed over three pints a day. Taking it by the gallon might answer in some cases for a short time, but could not be long continued.

DIET IN DIFFERENT DISEASES.

DIET FOR INFANTS.

RESORT to artificial food, though sometimes necessary, is always unfortunate for the baby. Trouble then begins. The baby fortunate enough to have a healthy mother and a natural diet acquires a strength and vigor which are of incalculable value in after-life.

For the first three days of the baby's life a little sweetened water in a spoon is all-sufficient. It is desirable to adopt, as far as practicable, regular periods for nursing. Once in every three hours during the daytime and about twice at night for the first month will generally be sufficient. After the first month three times during the day and once in the night will ordinarily suffice. This may be continued until the child is six months old.

According to many and the best authorities, no farinaceous food or thickening of any kind whatever should be given to a child under six months old. (See Appendix, page 216.) The child is until then "wholly unprovided with the physiological machinery requisite for the digestion of starchy foods." After six months the capacity for digesting starchy foods commences, and then a little gruel of sifted Graham flour, or barley, or cornmeal may be given. If the mother can nurse the child even partially, it is better to do so. If she cannot nurse the child at all, of course it is far better to procure a wet nurse than resort to cow's milk. Great care should be taken that the wet nurse be quite healthy and especially free from scrofulous or consumptive taint. If possible she should be of the same or nearly the same age as the mother, and her child and that of the mother should be of the same age. At nine months, or when the child has two or more teeth, it should be weaned. Not, however, during summer-time nor unless the child be quite well.

If the baby must be fed from the bottle the difficulties are many. The milk should be quite fresh from a healthy and properly fed cow. Then if the bottles are left to nurses to be cleaned there is constant danger that the work will be negligently or insufficiently done. It is absolutely essential that they be scalded and freed of all acid contents. The milk should also be given lukewarm, or near the temperature of mother's milk. Dr. Gatchel, in his admirable little book on "What Shall I Eat," says, "Half the sickness from which infants suffer is produced by improper food and improper feeding." Sir C. Clark, an eminent London physician, once said, "The ignorance of mothers in feeding children is worth a thousand pounds a year to me."

Cow's milk differs from human milk in that it contains more caseine, more butter, and more saline matter, but less water and less sugar. This difference must be rectified by adding to cow's milk the necessary water and sugar. For the first month give equal parts of milk and water; say of cow's milk one half-pint, of pure water (distilled or boiled) one half-pint, with powdered sugar of milk a teaspoonful or one lump of loaf sugar. If the child's stomach should be a little acid, a teaspoonful of lime-water can be added to this quantity.

After the infant is a month old use two parts of milk to one of water with sugar as above stated. The milk should be obtained fresh twice a day. Two pans should be kept exclusively for the baby's use, and, before the milk is added, the pans, used alternately, should have been thoroughly cleansed, scalded, and dried. In summer, the milk, if in danger of becoming sour, can be scalded when first put into the pan, but it must not be boiled. Glass jars are still better than tin pans for keeping milk.* Always use a fresh, clean bottle every time milk is given to the baby. Several bottles should be provided, also the black thimble rubber nipples; the white are said to contain injurious ingredients. Never use the long rubber tube for the nursing-bottle, as it is almost impossible to keep it clean and free from acid.

Dr. Gatchel says: "As soon as the child has taken enough for one feeding, empty from the bottle what remains, and, without delay, scald and wash the bottle with hot water and soap. After scalding, put the bottle into a basin of clean, cold water in which a little soda has been dissolved. Let it remain in the soda bath for half an hour, then rinse it in clean water and let it dry by hanging inverted on a peg." A wire basket would be better and more convenient than the peg.

While the baby is under a month old, the usual quantity for a meal should be the ordinary feeding-bottle half full. Afterwards the bottle nearly full.

In its chemical properties, goats' milk approaches nearer than any other kind to human milk. Very little water should be added to it—about four per cent. to make it suitable for infants.

Probably the most perfect artificial substitute for human milk is Liebig's food for infants, prepared according to strict chemical principles. It is composed of malt flour, wheat flour, cow's milk, bicarbonate of potash, and water in such proportions as to imitate woman's milk as nearly as possible.

---

* In regard to glass jars for keeping milk, see page 12.

Liebig's Receipt.

Take half an ounce of wheat flour, half an ounce of malt flour, and seven and a quarter grains of crystallized bicarbonate of potash, and, after mixing them well, add one ounce of water, then five ounces of cow's milk. Warm the mixture, continually stirring, over a very slow fire, till it becomes thick. Then remove the vessel from the fire, stir again for five minutes, put it back on the fire, finally let it boil well.

It is necessary that the food should form a thin and sweet liquid previous to its final boiling. Before using, it should be strained through a fine hair sieve.

Pavy says, in regard to this receipt: "To avoid the trouble of weighing, as much wheat flour as will lie on a tablespoon is an ounce, and a moderate tablespoonful of malt flour corresponds with half an ounce."

It is malt made from barley that should be used, and a common coffee-mill answers the purpose of grinding it into flour, which is to be cleaned from the husk by a coarse sieve. The bicarbonate of potash is added to neutralize the acid reaction of the two kinds of flour, and also to raise the amount of alkali in the food to the equivalent of that in woman's milk.

The ferment in the malt leads, during the exposure to the warmth employed in the process of preparation, to the conversion of the starch of both the flours into dextrine and sugar, the latter of which gives the sweet taste that is required. The newly found products also being soluble, will account for the mixture being thin, and the point contended for by Liebig is, that the same principles in this state tax the digestive and assimilative powers of the infant much less than starch.

Pap or Thickened Milk.

Ingredients: One pint of milk, two even tablespoonfuls of flour, and a teaspoonful of sugar. The sugar is often omitted.

Place the milk in a double boiler; when hot, stir in the flour, wet with two tablespoonfuls of cold milk; let the water in the outer vessel boil hard for an hour. Or, the pap can be cooked directly over the fire, when ten minutes' simmering will be sufficient to cook the flour. Proper care should be taken, though, to prevent scorching. This is pap proper; but for a change, when it is just done and hot, the white of an egg beaten to a stiff froth can be stirred in smoothly, without further cooking.

It is very desirable to use the new-process flour (in which the full amount of gluten is retained) or cold-blast flour prepared by the Health-food Company.

CRACKERS FOR BABY (over six months old).

Crackers may be fed to babies over six months old. Either the Boston or soda crackers, the health-food lactic wafers, or cold-blast biscuits, or crackers made at home (see p. 122), of new-process flour, can be used.

Pour over the cracker on a plate enough boiling water to cover it. Cover this with a saucer and let it remain in the oven for twenty minutes, or until it is quite soft and swollen. Then pour over some hot milk or thin cream.

BREAD JELLY, OR PANADA

is most excellent for babies ten or twelve month old. (See p. 215.)

GRUEL FOR BABIES.

Any of the gruels are good for the baby. The barley gruel is most excellent. If troubled with constipation a cornmeal gruel is generally better than medicine. If with summer complaint, the flour gruel or pap is advisable.

Oatmeal Gruel.—(Dr. Rice of Colorado.)

Oatmeal is a very hearty food, too much so to be commended as a common diet for infants. For a change, though, it often suits most admirably.

Add one teacupful of oatmeal to two quarts of boiling water very slightly salted; let this cook for two hours and a half, then strain it through a sieve. When cold, add to one gill of the gruel one gill of thin cream and one teaspoonful of sugar. To this quantity add one pint of boiling water, and it is ready for use.

Beef.—(Dr. Rice.)

Scrape one half pound of beef, and remove all the shreds; add one half pint of water, and three drops of muriatic acid. Let it stand one hour; then strain it through a sieve, and add a very small portion of salt.

Mellin's Food for Infants.

Mellin's food for infants, which is said to be merely the Liebig receipt carried out perfectly, is probably as good food for infants as can be purchased. It is an English preparation, but can be found for sale everywhere, as the food is well known and much used.

DYSPEPSIA.

Many conditions are requisite to insure good digestion, viz.: Wholesome food; food taken at proper intervals, so that it may be digested, and the stomach allowed some repose before another repast is taken; sufficient sleep; a mind free from nervous irritation, yet freely employed with projects, either useful or ornamental; a rejection of stimulating beverages, condiments, and spices; and, important as the selection of food itself, physical exercise. The working classes have the health and strength. Men, as a class, with their active vocations, are healthier than women. Fashionable women, as a general thing, do not take enough muscular exercise to keep themselves in good condition. Walking is all very well, but it is only about a third enough. The chest muscles, the liver, the vital organs generally, do not get enough stirring up, by bending, twisting, lifting, etc., to keep up a natural circulation, much less to create a healthy demand for food. The demand for food under normal circumstances is in proportion to the amount of organic expenditure.

It is this need of a free circulation of blood to take up food and carry it to perform its necessary functions, that causes half of the suffering from dyspepsia. Adirondack rowboats, mountain climbing, garden-making, and Dr. Oswald's woodshed matinées are all conducive to health and strength. Let those leading sedentary lives in the cities make Trianons of farmhouses and mountain camps, and play for a few weeks the *rôles* of English dairymaids and French peasants or American pioneers. Health and civilization would be advanced. One is the complement of the other—" *Les extrêmes se touchent.*"

The new gymnastic machines (the Gifford patent) are admirable, if it did not seem something of a waste of power to thrust and wave the arms in the air to no practical purpose. It is possible, however, that the same systematic action and the same exercise of particular muscles could not be accomplished so well in any other way. The exercise of sweeping and cleaning a room is admirable for bringing many muscles into play. Yet out-door work—horseback riding, garden making, snow shovelling, water pumping—is better and pleasanter. What a blessing it would be if all the hydrant water of the dwelling had to be pumped to the fourth story! and pumped by the proprietors of the castle.

Some do good service by exercising mornings and evenings in half undress with light dumb-bells. A happy idea is to shovel sand from one box to another, and continue the occupation daily. But if useful exercise which can interest the mind can be chosen, it is far preferable. If one wants to learn how to exercise, or rather to understand the importance of certain movements—it is well to do everything scientifically—a few lessons at an establishment where gymnastics, Swedish movements, exercises with vibratory-motion machines, and massage are conducted by competent physicians, would be of benefit. There are several such establishments in the large cities, and they are rapidly gaining in favor.

Another principal cause of dyspepsia is the general taking of too highly seasoned food. Then follow the drug poisons.

Dr. Oswald says, in relation to drugs: "What such tonics do is this, they goad the system into a transient and abnormal activity incident to the necessity of expelling a virulent poison. . . . The system has wasted the organic energy which it seemed to revive."

In chronic cases the best practice is undoubtedly to take all the out-door exercise possible, short of fatigue, to choose the most wholesome of foods, and patiently await results. "Temporary blue devils are far preferable to a persistent blue-pill Beelzebub."

Mais pardon! disciples of Esculapius. The author is going too far without a physician's certificate, and should only talk about diet.

Different foods must be tested, for what agrees with one, will not agree with all. A milk diet with farinaceous foods — oat-meal porridge, cracked wheat, corn bread, etc.—act like a charm with some, while a few persons cannot digest milk. Koumiss and peptonized milk can generally be relied on when simple milk is unsatisfactory. Raw-meat sandwiches, and the minced beefsteak (page 143), with as much pepper and salt accompaniment as can be dispensed with, is often beneficial; though meat should not be taken at the same time with milk. Baked potatoes, mashed, with cream, poached eggs, uncooked eggs (page 141), baked apples, and stewed fruits generally, are quite wholesome. A most important article of diet for dyspeptics is Graham bread made of wheat partly or wholly denuded of its fibrous coating.

A breakfast consisting of an oat-meal porridge, a cracked-wheat mould, or a generous slice of Boston brown bread, with cream poured over it, with hot water served in a teacup (see page 4) in place of tea or coffee—this and nothing more; a dinner composed of a slice of rare roasted or broiled beef, mutton, or venison, or a piece of well-cooked chicken, or broiled fish, for a change (only one of them at a time, however), one or two vegetables, and a rice pudding, a blanc-mange, custard, or other plain pudding; and a supper or luncheon of bread and milk, or cornmeal mush and milk, a mock-cream toast, and a baked apple, or some stewed fruit—this well cooked and lightly seasoned will generally appease an intractable stomach.

Let nothing be over-seasoned. Too much salt produces more or less inflammation and fever, and some hygienists banish it altogether, with the spices and condiments. They argue that food contains already enough salt. Mattieu Williams says: "Salt is not a condiment, but a food, simply because it supplies the blood with one of its normal and necessary constituents, chloride of sodium, without which we cannot live. A certain amount of it exists in most of our ordinary food, but not always sufficient."

It should probably be used much more sparingly than is customary.

Dietetic reforms should begin with a strictly nonstimulating diet. Let grape juice, koumiss, or currant-jelly water be the strongest beverage.

Salt or smoked meats, sausages, viands recooked, pickles, canned tomatoes, and fried dishes generally should be eschewed. And yet the diet must not be insipid. If well cooked and artistically served the admissible dishes would be relished by any one with a normal and healthy digestion — one not impaired or perverted by stimulants. The dietary suitable for a healthy child is generally suitable for an adult.

In extreme cases of irritability of the stomach, if milk, or milk and lime-water, koumiss or buttermilk, will not answer, the alternatives are barley water, the gruels of Graham flour, oat or corn meal, beef tea or oyster or clam broth. This is administered at regular intervals until the stomach evinces a more efficient working capacity.

Any dyspeptic may better undereat than overeat. A weak stomach must not be overtasked. Some physicians go so far as to say that total abstinence for a day or more, to give the organ a rest, is beneficial. If the dyspeptic could make up his mind to stop eating while still a little hungry, greater benefits would result than from the taking of nostrums.

It must be borne in mind, however, that while abstinence from food may be resorted to in special cases, dyspepsia can be brought on by fasting or by insufficient diet. The digestive functions can become weak from mere inertia. The tone of the stomach, like the tone of the muscles, may be lost by want of exercise.

Undoubtedly, as a rule, we eat too much. Persons of sedentary habits often eat as much as those employed in physical labor. Indeed, it may be noticed that the less one has to do, the more attention one gives to taxing the stomach. Not that one should scorn a good healthy appetite—but it is still carried to excess, especially by the world of people who lead sedentary lives, also by many wealthy families who consider that good living and hospitality require too great a variety of dishes, and too many courses at meals.

Dio Lewis, in one of his books, tells a story of a quack country doctor who advertised that he could cure any person of dyspepsia in a few weeks—price $400. The patient was sworn to secrecy as to the mode of cure, before being admitted to his sanatarium. His country patients expostulated, sometimes, in regard to the price. The doctor was obdurate, and as dyspeptics generally are hovering on the brink of despair, they invariably, sooner or later, came to terms. They were also invariably cured, according to the tradition. Finally the doctor died (not of dyspepsia, however), and one of his patients, considering his vow no longer obligatory, told the great secret.

A sanatary diet was of course administered; but the chief means of cure consisted simply in the patient kneading and beating his liver and stomach. At first it was sensitive and painful, but by careful rubbing and patting the exercise was daily increased, until the patient could pound the refractory organs vigorously for an hour or more at a time.

This may all seem extremely ridiculous. Not so, however. It is really the Swedish movement-cure to a new tune—a *pas seul*. The soreness of even a boil can be reduced and sometimes removed by careful manipulation. The circulation is thus equalized, giving new strength, and carrying off the poison.

DIARRHŒA, DYSENTERY, AND CHOLERA.

Diarrhœa.

Diarrhœa results from an effort of nature to throw off either an excessive quantity or a poor quality of food which cannot be digested. The digestive powers in such cases are overtaxed and weakened, and the best remedy in the first stage of an acute attack is total abstinence from food for at least a day. The stomach needs rest, and the patient will not suffer from this fasting, but will often recover by simply retaining a recumbent position and taking nothing but a little cool water, or, at most, rice-water, in small quantities at a time. For the first two or three days a little rice gruel will be sufficient in the way of food. If milk agrees with one perfectly, it can be taken mixed with lime-water (a tablespoonful of lime-water to a gobletful [half a pint] of milk), at first at intervals of one or two hours. After a time, as strength is developed, the quantity may be increased to, say, a small glassful every three or four hours. Milk is generally an excellent diet for this trouble, and, when taken, nothing else is required. Koumiss (new or freshly made), is also highly recommended for diarrhœa. Thickened milk or flour gruel is often given. There are some who cannot take milk, and then the alternatives are barley-water, thin oatmeal gruel (strained), beef tea, oyster broth, and sometimes the pulp of raw meat.

The patient should be extremely careful during convalescence to take only the most digestible of foods—for instance, toast dipped in milk, raw egg (page 141), rice puddings, tea and toast sippets (soaked in tea), the preparations from the health foods, etc.

DYSENTERY.

In this disease there are inflammation and ulceration of the intestines. Consequently the patient should be kept in as tranquil a state as practicable. The food should be such as to exert the least stimulating or irritant action on the mucous membranes. An exclusive diet of milk (given as described in the preceding article), is of as great value in dysentery as in diarrhœa. Rice-water and rice gruel are also especially recommended, as well as barley and flour gruel.

Raw eggs (page 141), or eggs lightly poached, or eggs beaten with milk and sweetened, as described for milk punch (without the liquor), are useful in dysentery. The pulp of raw meat is sometimes advantageously used in cases of diarrhœa and dysentery. The fat is all removed and the fresh beef is either scraped and divested of all fibre, or it can be cut into a pulp with a mincing-machine. This fine pulp may be lightly seasoned with salt and red pepper and placed between two thin slices of stale bread, forming a sandwich; or it can be formed into a thick cake and the outside merely colored by placing it in a hot saucepan; but the inside must not be cooked.

Dr. Hall gives a tablespoonful of scraped raw beef every four hours.

Cold drinks tend to aggravate the pain and colic which accompany this disease.

CHOLERA.

During the prevalence of cholera great care must be

taken to keep digestion in good order. No ice-water, alcoholic stimulants, stale or unripe vegetables, pickles, or any indigestible food should be taken.

Dr. Gatchell says: "During the attack no food whatever is required. The incessant thirst from which the patient suffers it is hard to gratify, for water taken into the stomach aggravates the vomiting; and yet the patient should receive all the water that he craves, *if he can retain it*. If this is impossible, much benefit may be derived from holding small pieces of ice in the mouth until they melt away. Injections of warm milk may be used with advantage, if nothing can be taken by the stomach.

"After the attack no solid food should be taken until the stools are consistent and fæcal. Great care must be observed during convalescence. An attack of indigestion at this time is often followed by a fatal relapse. At first only farinaceous food should be given, and this in small quantities, frequently repeated.

"Rice thoroughly cooked, thickened milk, and the like may first be taken. Milk, however, is to be preferred to this, and, if the patient can take it, nothing else need be sought for."

FEVERS.

Dr. Beaumont found, by experimenting with a young man who had his stomach opened by a musket shot, and afterwards so covered that the action of the gastric juice could be witnessed, that but little gastric juice is secreted in febrile diseases.

The digestive power is very weak. Fevers seem to be due to a poison multiplying itself in the blood, which runs a regular course, more or less severe according to the different constitutions attacked and according to the nursing and care received. The body becomes emaciated. Both the tissue and adipose matter seem to burn up. Cooling drinks and food only in liquid form should be given, and the latter in small quantities, but at regular intervals of, say, two or three hours. Solid food given even during convalescence will often cause a relapse.

In some stages of fever there is an intense longing on the part of the patient for cool air, cold water, and especially for acid drinks, and but little desire for solid food. All the pure cold water that is desired should be given. Barley and toast water can be given also as drinks. Lemonade, orangeade, tamarind and currant-jelly water, and buttermilk are generally craved, and, if so, they are beneficial. Milk fresh from the cow, or else ice cold, as preferred, is recommended by all the authorities for fever patients (except in cases of typhoid fever). Koumiss is especially beneficial for fevers. It is always received gratefully, and is the very best diet that can be given, as it contains a mild stimulant in addition to its digestible food properties. Beef tea and koumiss or milk can be given in alternation. Barley, oatmeal, and Graham-flour gruels are much used, especially during convalescence. It is well not to use stimulants unless the patient is alarmingly weak, when an eggnog can be given if koumiss is not a sufficient stimulus.

Peptonized milk and gruel are also recommended when fresh milk and gruel do not agree with the patient. If milk disagrees, or is thrown up curdled, a tablespoonful of lime-water to a cupful of milk will generally correct the difficulty.

The fruits are especially beneficial to fever patients. Dr. Oswald says: " Bananas are *par excellence* an anti-fever food, being refreshing, palatable, and nutritive, as well as exceedingly digestible."

When the patient has no appetite for food, very little or none should be given. Dr. Oswald says: "When coolness, sweetness, and fruity flavors cannot make a dish acceptable to the appetite, its obtrusion would do more harm than good, and it is a great mistake to suppose that even total abstinence could, in such cases, aggravate the danger of the disease."

In the critical stage of fevers, milk, koumiss, a light gruel, orange juice, and the cold drinks are all that need be given. After the crisis has passed, bananas, pears, baked apples, raw eggs (page 141), bread jelly, dipped toast (made of nutritive flour), or barley gruel could be added. All animal food or greasy dishes should be avoided until full recovery.

If the patient's mouth is furred it may be washed out with cold water containing a little lemon juice, before food is taken.

Typhoid Fever.

This being a long and exhausting disease, the chief treatment consists in good nursing and careful diet. In this disease the lining membrane of the intestines becomes ulcerated. This complicates the question of diet, as nothing should be given which will leave a solid residue in the bowels, for fear of irritating the ulcers and causing them to perforate through the intestines. This cuts off fresh milk, the curd of which forms more or less solid masses. Koumiss which is partly digested is to be preferred to milk, also peptonized milk gruel (see pp. 42, 213, 228). Beef tea is also beneficially given.

The experience of Sir Wm. Jenner is so extended in the treatment of typhoid fevers that I add his remarks on "Diet in Typhoid Fever" in the Appendix (page 223).

There is rapid waste in this fever, and the patient must be fed regularly with very nutritious food. Koumiss, beef tea, the gruels, eggnog, etc., are the chief articles of diet. If the patient becomes unable to swallow, nutrient enemeta must be resorted to. Rubbing the body with oil is of great value.

GOUT AND RHEUMATISM.

An excess of uric acid in the system, and the consequent tendency to deposit urate of soda in the fibrous tissue around the joints, is the cause of gout. This condition is superinduced by the use of too much highly seasoned animal food and by indulgence in stimulants, without taking sufficient physical exercise. In other words, more food is taken than can be properly digested and assimilated. Laborers, taking a less proportion of animal food, and more out-door exercise, are rarely ever troubled with gout. Gouty patients and the children of gouty parents should promptly adopt habits of strict abstemiousness. The diet should be chiefly vegetable, and physical exercise in the open air is indispensable. The regimen recommended for dyspepsia will answer very well for gout and rheumatism—rheumatism, like gout, being often consequent upon dietetic abuses. Meat should be strictly avoided in all cases of gout. It will only aggravate the trouble, and the same may be said of alcoholic drinks, malt liquors, and especially port wine.

Probably the surest cure for both gout and rheumatism is to be found in a complete change of the ordinary dietary in favor of a purely milk diet—or a diet composed of milk and the grain preparations, viz.: oatmeal porridge, cracked wheat, Graham bread, etc. Skimmed milk is generally prescribed, but the patient must be ill indeed if not able to digest fresh, new milk.

The patient need not fear starvation. He will find himself stronger than ever. Milk contains all the elements necessary in food, and it contains these elements in the proper proportions to promote digestion and to produce healthful assimilation.

Probably one would not be obliged to continue this strict dietary for more than a few weeks, before a more varied *menu* could be trusted.

BRIGHT'S DISEASE.

Pavy says: "Physiology teaches us that the kidneys perform an eliminative office. The water which they remove in regulating the amount of fluid in the system is made the vehicle for carrying off solid matter, consisting of useless products of the metamorphosis of the food, and effete materials, resulting from the disintegration of the tissues, which poison and produce death, if allowed to accumulate in the blood. In Bright's disease their eliminative capacity is interfered with.

"The amount of urinary matter to be discharged is largely dependent upon the nature of the food. The fats and carbohydrates* throw no work upon the kidneys. The products of their utilization—carbonic acid and water—pass off through another channel.

"The nitrogenous ingesta, on the other hand, in great part undergo metamorphosis, and yield their nitrogen to be carried off in combination with a portion of their other elements, under the form of urinary products. In this way the kidneys become taxed by the food. So a vegetable diet should preponderate.

"It must not be lost sight of, that, on account of the escape of albumen, an extra amount of nitrogenous matter should be supplied to make up for the loss of albumen. In Bright's disease the kidney is contracted, and frequently the escape of albumen is insignificant, and

---

* Composed of starch.

sometimes even it is none. The mere loss of albumen is not so much to be dreaded as uræmia."

A vegetable diet is also recommended by most of the authorities (Chambers being an exception), on the supposition that meat throws extra work upon the kidneys. In the use of the grain foods such preparations only should be selected as contain the full nutriment of the grain (see pages, 26, 207).

A diet wholly or partially of milk is much recommended. Niemeyer says: "In a series of cases which have been described by Dr. Schmidt, in his inaugural thesis, I have obtained most brilliant results, where all other treatment has failed, by putting the patients on an almost exclusive diet of milk."

The ordinary mixed diet should be gradually changed in favor of the milk diet, until one exclusively of milk is finally reached. This should be kept up for a month or so, when improvement is almost certain.

The patient should drink freely of pure soft water, as that carries off much of the impurities of the blood. Flax-seed tea is at times beneficial. No alcoholic or malt liquors should be allowed in any form. They act as a certain poison in kidney affections, and their excessive use, without doubt, is the provoking cause of a majority of such diseases.

DIABETES.

The formation of sugar in the urine is what is characterized as diabetes. The cure of this disease is almost entirely dietetic, and consists merely of the patient and persistent taking of foods which contain no sugar nor starch, which latter is converted by natural processes into sugar in the system. Fat and albuminoids are given in their place. Dr. Dobell recommends very highly the pancreatic emulsion of fat for diabetics.

The following is a list of dishes which are allowed and prohibited a sufferer from diabetes. The dishes are allowed which are not marked prohibited.

Oysters and Clams.

Raw or cooked without flour mixtures. Oysters can be rolled in egg and gluten for frying.

Soups.

All kinds without flour, rice, or other starchy substances, and without the prohibited vegetables.

Fish.

All kinds, including lobsters, crabs, sardines in oil, etc.

Meats.

Of all kinds. Poultry, game, etc. Livers, on theoretical grounds, are prohibited.

Vegetables Allowed.

Cauliflower, spinach, cabbage, string beans, cucumbers, lettuce, greens, mushrooms, young onions, and olives. Celery, asparagus, and tomatoes are questionable. Sour apples cut in quarters, dipped in egg and rolled in gluten, and fried in hot fat, make a good substitute for potatoes, and may be used moderately.

Vegetables Prohibited.

Potatoes, beets, turnips, pease, beans, carrots, parsnips, rice, sago, tapioca, vermicelli, or others containing sugar or starch.

Milk, Cheese, and Eggs.

Milk, in some cases; eggs, cream, butter, buttermilk, and all kinds of cheese may be taken freely. Puddings and custards should be sweetened with glycerine.

Fruits.

All kinds of tart fruits, peaches and strawberries with cream and no sugar.

Fruits Prohibited.

All the sweet fruits, as apples, pears, plums, grapes, bananas, pineapples, raspberries, blackberries, etc.

Breads and Pastry.

Only those made from wheat-gluten flour. The ordinary flour or grains (oatmeal, cornmeal, hominy, etc.) must not be used in any form.

Beverages.

Koumiss, coffee with cream and glycerine (no sugar). Cereal coffee, very good. Tea objectionable. No liquors nor wines, except claret, Rhine, or other acid varieties. It is still better to reject all wines, sweet or sour, and all liquors, malt or distilled. As much pure water as desired may be taken.

Nuts Generally.

Plenty of exercise in the open air, tepid baths, rubbing, and abundant sleep are desirable.

CONSUMPTION.

The principal object in treating consumption is to build, the tendency of the disease being to waste. There must be, if possible, a renewed and healthy organic growth to arrest the formation of tubercular and diseased matter; consequently all the nourishing food which can be digested and assimilated should be taken. Plenty of fresh milk, if possible warm from the cow, is desirable; also buttermilk, clabbered milk, and koumiss (see articles on Koumiss).

Fresh meats, such as beef, mutton, and venison, roasted or broiled, and cooked rare, should be freely indulged in (meats and fresh milk must not be taken at the same meals, however). Fowls and fresh fish may be safely and profitably taken. Pork, veal, and all foods difficult and slow of digestion should be avoided. All salted meats should be eschewed. Potatoes, carrots, and fresh vegetables generally, are wholesome, and even necessary, when much meat is taken. Raw and slightly cooked eggs are full of nutrition and very assimilable. Care should be taken to discontinue at once any article of food that disagrees with the patient, as disordered digestion is especially unfortunate in consumption.

As much fat as can be digested, whether it be in the form of cream, butter, fat of meat, or oil, should be taken. Cod-liver oil seems to be one of the great resources for supplying fat to consumptives, and the amount of evidence accumulated in its favor leaves no doubt as to its utility. The oil should be quite fresh, without color, and should be kept well corked in a cool place. If it does not agree in its crude form, there are preparations of it in emulsion, combined with pancreatic extract, malt, hypophosphates, etc., which are considered beneficial and should be tried.

Dr. Gatchell says a dose of a teaspoonful of cod-liver oil is sufficient to begin with, and this quantity can be increased until a tablespoonful three times a day may be safely and profitably taken. It must not be taken on an empty stomach, but half an hour after a meal.

The pancreatic emulsion (see pages 40 and 213), a preparation of half-digested beef suet, is well worth a trial.

Alcoholic stimulants are considered very injurious to consumptives by most authorities, and they should not be used at all except in hopeless cases, where they may serve to give temporary strength in periods of extreme weakness or to alleviate acute pain and suffering. The effect of alcohol in any but the smallest quantities (as found in koumiss, etc.), is to derange and weaken the digestive powers, the main reliance for cure.

Among others, Dr. Chambers says: "As to the use of alcohol in threatened cases, and in the early stage of tubercle, I have no hesitation in pronouncing an opinion against it."

As nothing aids digestion, and consequently assimilation and health, so much as fresh air and sunshine, combined with all the physical exercise that can be borne *without fatigue*, a life in the mountains, where the air is dry and bracing, is to be chosen if possible. Having spent three summers in the mountains of Colorado, and having seen and conversed with many consumptives, I am led to believe that the cures are in almost all cases among those who adopt an out-door tent life, and impose upon themselves a certain amount of physical work. I say *work* purposely, to designate something more than mere exercise. Actual work giving good exercise to the arms and chest is especially desirable, always remembering to stop short of fatigue.

The most remarkable cure of which I knew was that of a man far gone with consumption (as they avowed), and hardly able to walk when he started from Missouri, who made the trip across the plains in an ambulance, and soon cooked his own and companions' meals. The trouble with most invalids is that they haven't "vim" enough to be willing to work for health. In the Adirondacks of northern New York, among the hemlocks "on the Raquette," a long distance from any first-class hotels (which are all very well in their place), we have met many consumptives, and in all cases they were benefited by the wild-woods life. Some persons spend the winter there and take their "constitutional" by chopping wood, etc., and report that the winters are even more beneficial than the summers.

Dr. Chambers says: "The use of climate in the treatment of phthisis (consumption) may be tested by its dietetic action; if it improves the appetite it is doing good; if it injures the appetite it is doing harm."

Scrofula.

The diet in scrofula should be the same as in consumption; a full diet containing plenty of fat, in the way of cream, fresh milk, butter, fresh animal food, cod-liver oil, etc., and also a full complement of fresh air, sunshine, and exercise. The extract of malt is generally recommended.

Rickets.

This disease is the result of imperfect nutrition, and should be treated like scrofula, by prescribing a gencrous diet, such as milk, cream, raw beef, and cod-liver oil. The extract of malt, which contains phosphates of lime and other salts, is especially valuable in the treatment of this disease.

DIPHTHERIA.

The patient should be well nourished. Give plenty of fresh, new milk, or milk mixed with beaten egg (milk punch without the liquor).

In the stage of depression some stimulant is required. Let it be eggnog, milk punch, or raw egg beaten with a spoonful of whiskey or brandy, oatmeal caudle, or koumiss.

If the patient can no longer swallow, he should be nourished by nutrient enemeta, and by rubbing the body, especially the abdomen (under cover, for fear of taking cold), several times a day with olive-oil.

GASTRITIS.

In the height of the attack, when the stomach is much inflamed, no food whatever should be taken. Small pieces of ice may be held in the mouth and some swallowed. Fresh koumiss is most excellent. Ice-cream flavored with lemon extract (no vanilla) is also valuable. If milk agrees, no other food is required. The gruels come next, but no meats should be eaten. The meats are digested in the stomach, and the starchy foods in the large intestine beyond the stomach.

If nothing can be retained on the stomach, nutrient enemeta and rubbing the body with oil must be resorted to.

CORPULENCY.

Fat in the body is created out of the fat of food, and also from its starch and saccharine elements. Consequently, in the treatment of corpulency, it is necessary to interdict foods that contain fat, starch, or sugar. Sugar, according to Banting, is the most active of fat-forming foods.

Mr. Banting's rules were as follows:

"For breakfast, at 9 A. M., I take five or six ounces of beef, mutton, kidneys, broiled fish, or cold meat of any kind except pork and veal; a large cup of tea or coffee without milk or sugar; a little biscuit, or one ounce of dry toast.

"For dinner, at 2 P. M., five or six ounces of any kind of fish except salmon, herring, or eels; any meat except pork or veal; any vegetables except potato, parsnip, beet, turnip, or carrot; one ounce of dry toast; fruit out of a pudding not sweetened; any kind of poultry or game, and three or four glasses of good claret, sherry, or madeira (champagne, port, and beer forbidden).

"For tea, at 6 P. M., two or three ounces of cooked fruit, a rusk or two, and a cup of tea without milk or sugar.

"For supper, at 9 P. M., three or four ounces of meat or fish with a glass or two of claret, or sherry and water."

The propriety of the last meal, or of the taking of

sherry or madeira (heat-producing wines), or of rusks, which are sweet biscuits, is doubtful.

The following comprise the fat-producing foods, viz.:
Milk, cream, butter, fats, soups, puddings, pastry, sugar, candies, cake, and all sweet dishes, rice, corn-starch, and all the farinaceous foods (excepting toasted bread or bread crust), potatoes, corn—in fact all edible roots and vegetables growing under ground—sweet fruits, and spirituous and malt liquors.

The following are non-fat-producing foods, viz.:
All the meats, poultry, and game, with the exception of the fat portions thereof, oysters and shell-fish; celery, spinach, and all the greens, cabbage, onions, lettuce, squash, tomatoes, and other vegetables containing little or no starch, and all acid fruits.

Dr. Dobell thinks that a certain amount of fat should be taken with the food. On this subject, he says: "On comparing the following analysis of Mr. Banting's diet for getting thin with my tables of normal diets, it will be seen that it yields less than half the normal quantity of carbon, leaving the deficiency to be obtained from the fat already stored up in the system, by the consumption of which the obesity is removed. The fault consists in this reduction of carbon being obtained by diminishing the hydrocarbons (fats) of the foods instead of only cutting off the carbohydrates (sugar and starch). It has happened to me to have much to do with a great number of persons who have tried Bantingism, and I do not hesitate to say that Mr. Banting has done a great deal more harm than good. Mr. Banting candidly told his readers that he was ignorant of the physiology of food.

"The facts in the case are these: 1. A certain amount of fat in the system is one of the most essential elements of health. 2. The quantity required by different individuals to maintain health differs. As much fat should be taken as the stomach likes. 3. The effects of a deficiency of the quantity actually required are most disastrous, the tissues of the body and the brain and nerves being at length disintegrated to supply the elements of fat which they contain. 4. When there is a quantity of fat in the body in excess of that necessary to health, it may be lessened with great and (needed) advantage, provided it be done slowly and without cutting off too much of the fat element of food."

There is much to be gained by observing certain other rules, aside from the dietary. For instance, every morning a hasty cold water sponge-bath should be taken, and the body should be well rubbed with a crash towel. And whenever the body is too warm, the cold water sponge-bath may be repeated without a general undress. The clothing should not be too warm. All the bodily exercise that can be taken without fatigue should be persistently kept up. The vibratory-motion machine is most excellent for reducing fat. This machine makes two thousand vibrations a minute, and is made to accommodate different portions of the body. To those who are unable to take other and ordinary exercise this machine is especially recommended. Its action is to produce a rapid circulation of the blood, which takes up and carries off adipose and effete matter.

With plenty of exercise there will be less need of an exacting dietary.

SOMETHING ABOUT LONGEVITY.

Before the age of eighty, it is not years that make us old. It is want of health, either inherited or brought on by our own imprudences. Health is youth. Many are younger at sixty than others at twenty. The person in health is always young. The invalid is always old. To him life is without enjoyment, without energy, and without aspiration. And yet, when health is everything, life itself, how little it is guarded! how little appreciated, except when lost! What a plaything it is! And so our youth is our middle age. Our middle age is our old age. When it is time to live and enjoy the fruits of experience, study, and labor, we are practically dead. Men have lost their vigor at sixty, and women have lost their beauty at forty.

There are some who stop to think. They discuss ill-ventilated bedrooms, temperance, corsets, graham bread. *Dieu nous défend!* What disagreeable subjects! What cranks and crazy theorists they are! These theorists attack established habits. Fixed habits are tyrants, and their power is irresistible, and so the study of health is unpopular. Sickness alone must be considered.

There is no doubt that the natural period of human life is greatly shortened by long and perverse violation of natural laws, and that the requirements to guard the divine gift are many.

The total length of life among dumb animals is about five times the period between birth and full maturity.

A horse is mature at from five to six years. His average age is five times as much, and so the rule holds true with the other animals. Man is mature at twenty-five. On the same principle he should live to be one hundred and twenty-five. This is a charming world, and the author cannot afford to make the mature age at less than twenty-five, especially when man is not considered sufficiently aged to hold most of the important offices before thirty.

At least, a lesson can be learned from animal life. Animal food is simple. It is without spices. The drink of animals is water. Their bedrooms are ventilated. They breathe pure air. Bad colds and dyspepsia are infrequent. They inherit sound constitutions.

Climate has much to do with the preservation of youth, or rather health. In the extended territory of the United States many healthful situations are to be found; not, however, where the weather is very changeable, nor where the average temperature is very high. In some portions of Scotland men often retain their full vigor at eighty. The equable climate of England is especially salubrious.

Reference to a few examples of persons living to a great age may be both interesting and profitable.

Cornaro, a distinguished Italian nobleman, found himself at forty quite broken down by his gross excesses. Upon the advice of a physician he resolved to lead a new life—to maintain a temperance which should be as marked as his former indulgences. At eighty-three he wrote a work, "Sure and Certain Method of Attaining a Long and Healthy Life." This work was followed by three others, written at the ages of eighty-six, ninety-one, and ninety-five. His works were translated into Latin, French, German, and English. The English translation reached its thirty-ninth edition in 1845. Cornaro exclaims: "O blessed temperance, how worthy art thou of our highest esteem! and how infinitely art thou preferable to the irregular and disorderly life! There is as wide a difference between you as there is between light and darkness, heaven and hell." Again, he says, in older age: "O sacred and most beautiful temperance! how greatly am I indebted to thee for rescuing me from such fatal delusions, and for bringing me to the enjoyment of so many felicities, and which over and above these favors conferred on your old man, has so strengthened his stomach that he has now a better relish for his dry bread than he had formerly for the most exquisite dainties! My spirits are not injured by what I eat, they are only revived and supported by it." To a distinguished archbishop he again wrote: "Is it not a charming thing that I am able to tell you that my health and strength are in so excellent a state? that instead of diminishing with my age, they seem to increase as I grow old? all of my acquaintances are surprised at it, and I, who know the cause of this singular happiness, do everywhere declare it.... I confess it was not without great work that I abandoned my luxurious way of life." When Cornaro was ninety-five, he wrote: "I find myself as healthy and brisk as if I were but twenty-five. Most of your old men have scarce arrived at sixty before they find themselves loaded with infirmities. They are melancholy, unhealthful, always full of dreadful apprehension of dying."

There are many famous instances of longevity.

Count Jean Frederick de Waldeck died in Paris, in 1875, at the age of one hundred and nine. He had been conspicuously before the world for over ninety years. He became member and honorary member of the principal learned societies of London and Paris.

"The Irish Countess of Desmond fell from a fruit tree, broke her thigh, and died in 1609, aged one hundred and forty-five years. She danced at court with the Duke of Gloucester, afterwards Richard the Third. She continued gay and lively in her tastes, dancing even beyond her hundredth birthday. She cut three new sets of teeth."

The Cardinal de Salis, archbishop of Seville, who lived to be one hundred and ten, considered his health and vigor were owing to his care in diet.

But examples of longevity are generally found among the poorer classes. Lord Bacon, in his "History of Life and Death," thus quotes from Pliny: "The year of our Lord 76 is memorable; for in that year there was a taxing of the people by Vespasian; from which it appears that in the part of Italy lying between the Appenines and the river Po there were found fifty-four persons one hundred years old; fifty-seven, one hundred and ten years; two, one hundred and twenty years; four, one hundred and thirty years; four, one hundred and thirty-five years; and three, one hundred and forty years each."

Mr. Eugene Thompson, in an interesting article on "Longevity," published in *Scribner's Magazine*, in 1875, writes: "Now leave sunny Italy and go to inclement Norway. Travellers have there remarked the great temperance, industry, and morality of the people, and their common food is found to be milk, cheese, dried or salt fish, no meat, and oat bread, baked in cakes. An enumeration of the inhabitants of Aggershaus, in Norway, in 1763, showed that one hundred and fifty couples had been married over eighty years; consequently the greater number were aged one hundred or more; seventy couples had been married over ninety years, which would place their ages at about one hundred and ten;

twelve couples had been married from one hundred to one hundred and five years, and another couple one hundred and ten years, so that this last pair were doubtless one hundred and thirty years old."

Thomas Parr, buried in Westminster Abbey, died in 1655, at the age of one hundred and fifty-two. He lived in Shropshire, England, a place noted for its long-lived people. He was a farmer of extremely abstemious habits, his diet being chiefly milk and coarse bread. He married his second wife when one hundred and twenty-two, and worked at the age of one hundred and thirty. He was taken to court in his one hundred and fifty-second year as a curiosity, by the Earl of Arundel, and his life was prematurely cut off on account of the change from a parsimonious to a plentiful diet. Two of his grandsons lived to be each one hundred and twenty-seven years old, and a third grandson to be one hundred and nine; and Robert Parr, a great grandson, died in Shropshire, in 1757, aged one hundred and twenty-four.

Mr. Ephraim Pratt, of Shutesbury, Mass., who died at the age of one hundred and seventeen years, lived chiefly on milk, and his son, Michael Pratt, attained the age of one hundred and three by similar means.

A study of the subject shows us that great longevity has always been accompanied by abstemiousness in diet; also, that great eaters never live long.

UTENSILS.

A Double Tin Steamer with double tin cover and copper bottom is invaluable among cooking utensils, especially for making several dishes suitable for the sick. The double tight-fitting cover, perfectly securing the heat, cannot be satisfactorily supplied with any improvised cover. The steamer is also a valuable utensil as a *bain marie;* i. e., for keeping any cooked dish hot.

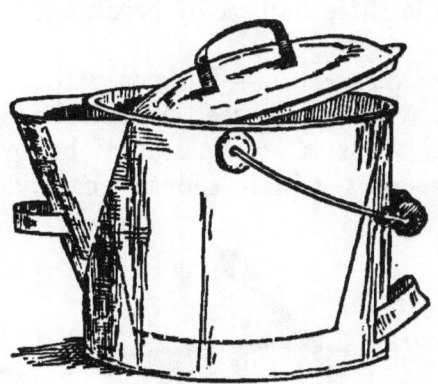

DOUBLE TIN STEAMER.

For this purpose the steamer containing hot (not boiling) water is kept at the back of the range. The double cover and the hot-water lining protect soup, vegetables, sauce, oysters, or any dish placed inside. The flavor of a dish is almost perfectly preserved when kept in this manner.

This steamer is especially useful for making Boston brown-bread, Graham pudding, farina pudding, custards, etc.

The Earthen Crock (see page 129) is recommended for cooking grains (oat-meal, etc.), apple sauce, the fruit compotes, etc. This crock must be heated gradually, when there is little danger of breaking.

A Copper Saucepan. — This is rather an expensive utensil, but when once used it will be considered indispensable. This is on account of being able to cook with it materials which scorch readily, viz., articles

with milk, cracked wheat, or any of the grains, sauces, etc., which are improved by simmering, with almost no danger of burning. The same materials could be cooked in a new porcelain kettle or earthen crock; but iron or tin saucepans—in fact, any kind—do not preserve the same even, regular heat as those made of copper. As porcelain kettles are not durable, the copper saucepans at last are cheaper. They will last forever. However, special care should be taken, if the copper is exposed inside, to have them at once retinned.

Meat-juice Press—for extracting the juice from meat. The meat—a thick steak cut from the round of beef preferable—is broiled merely enough to become well heated through. It is then cut into pieces an inch or less square, and put into the press, which has been previously heated by inserting both cup and cover into hot water. Juicy meat will yield nearly half its weight in liquid. An equal quantity of warm water is often added to the meat juice, and all should be very lightly seasoned. It can be reheated before giving it to the patient, although it should not reach the boiling-point, for reasons explained on page 100.

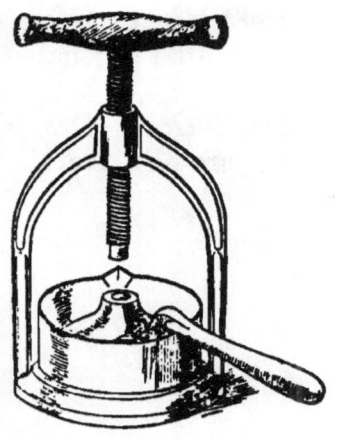

Porcelain Duck for administering drinks and fluid foods to a patient in a recumbent position. The narrow neck prevents a too rapid flow of fluid into the mouth.

The duck should be warmed before hot foods are poured in.

Glass Tubes come for the same purpose, which are also very convenient.

Porcelain or *Glass Spoons*, for administering medicines, can be purchased of any druggist.

Little Glass Droppers, for measuring medicine by drops, are also useful.

RECEIPTS
FOR THE SICK AND CONVALESCENT.

DRINKS.

DISTILLED WATER (Dr. Beard).

"For diseases of kidneys, etc., this, the purest of water, may be obtained by fixing a curved tin tube three or four feet long to the spout of a tea-kettle, and conducting its free end into a jar which should be placed in a basin of cold water. The liquid, as it drops, must be kept cool by frequently changing the water in which the jar is placed. The steam thus condensed is pure water. Distilled water is mawkish to taste, but this is easily corrected by pouring it from one vessel to another successively for ten or fifteen minutes, so as to involve in it a quantity of atmospheric air."

LIME-WATER.

Pour over a piece of fresh unslacked lime, about an inch square, two quarts of hot water. When it has slacked (in a few minutes) stir it thoroughly. Let it remain over-night. Bottle carefully all the liquid that can be poured off in a perfectly clear state.

As water will only hold a certain amount of lime in solution, the addition of more lime would make the water of no greater strength.

Lime-water (an alkali) is generally added to milk for the purpose of neutralizing the effects of an acid stomach.

About a teaspoonful to a tablespoonful of lime-water to a half-pint of milk is usually prescribed.

BARLEY-WATER.

Add to a pint of boiling water half a tablespoonful (half an ounce) of Robinson's patent barley, or the "health-food" barley, rubbed smooth, with two tablespoonfuls of cold water; add also a pinch of salt and a tablespoonful of sugar. Let it boil five minutes. It is to be drunk cold. The simple barley-water has a not unpleasant taste, and is often prepared without additional flavor. Yet zest—*i. e.*, the thin yellow cuts of the rind of a lemon, or lump sugar rubbed over to extract the oil—can be added as a flavoring, or a lemonade may be made of barley-water.

Barley-water may be used temporarily instead of milk when the latter disagrees.

OATMEAL DRINK.

Rub two tablespoonfuls (two ounces) of oatmeal smooth by gradually stirring in a teacupful of cold water; add a pinch of salt. Stir this into a quart of boiling water and let it boil half an hour. Strain it through a fine sieve.

TAMARIND-WATER.

Stir into a glassful of water a tablespoonful of preserved tamarinds.

CINNAMON-WATER.

Add five or six sticks (half an ounce) of cinnamon to a pint of boiling water, and boil fifteen minutes. To be administered by the tablespoonful.

Given for hemorrhages.

TOAST-WATER.

Toast thoroughly thin slices of Graham bread, and break them into a bowl. Pour over enough boiling water to cover it. When cold, strain off the water and sweeten it slightly. Serve it always freshly made.

CURRANT-JELLY WATER (for fever patients).

A teaspoonful of currant-jelly dissolved in a goblet of water, and sweetened to taste, affords a refreshing drink for invalids.

FLAXSEED TEA.

Add half a cupful of flaxseed to four cupfuls, or a quart, of boiling water. Let it boil half an hour. Let it stand fifteen or twenty minutes near the fire, after it has boiled. Of course the longer it stands the thicker it becomes. Strain, sweeten to taste, and add a little lemon-juice, or not, as preferred.

This is a useful demulcent drink for coughs, etc.

FLAXSEED AND LICORICE TEA (for coughs, etc.).

Pour one pint of boiling water over one ounce of flaxseed, not bruised, and two drachms of licorice-root bruised, and place the covered vessel near the fire for four hours. Strain it through a sieve.

HERB TEAS

are made by pouring boiling water over one or two teaspoonfuls of the herbs; then, after covering well the tin cup or bowl, allowing it to steep for several minutes by the side of the fire. The tea is then poured off, and sweetened to taste. Camomile tea is used for nervousness and sleeplessness; calamus tea, for infant's colic; cinnamon tea, for hemorrhages; watermelon-seed tea, for strangury, etc.

Wine, Lemon, or Vinegar Whey.

When a pint of milk is brought just to a boil, pour in a gill of sherry wine. Let it again come to a boil. When the whey separates, strain and sweeten it to taste, using perhaps a teaspoonful of sugar.

Or the whey can be made in the same manner with lemon-juice (free from seeds), using the juice of half a lemon instead of the wine, and sweetening to taste; or with vinegar, a tablespoonful being quite enough for a pint of milk.

In an alimentary point of view, whey is almost of no value. It is advantageous as a drink in febrile diseases, and is a good means of administering wine in small quantities.

It may be drunk either cold or warm. It possesses sudorific and diuretic properties.

Sugar Syrup (for sweetening drinks).

For drinks of all kinds, even tea and coffee, sugar syrup gives a better flavor than crude sugar.

To a cupful of white sugar add a quarter of a cupful of water, and let it boil one or two minutes. It must not be boiled long enough to candy. This syrup is also purer and better than most of those purchased, to eat with pancakes, etc.

Simple Beverages from Fruits.

Currant-jelly water (or any acid jelly — cranberry, plum, etc.).

If the jelly is soft a teaspoonful is dissolved in a goblet of fresh cold water, and sweetened to taste.

If the jelly is hard, it will have to be added to boiling water to become dissolved. To be drunk cold.

The fresh fruits are, of course, to be preferred.

There is nothing more refreshing than currant-water made from fresh currants. This can be prepared by allowing a pint of water to a pint of currants (freed from the stems) and a tablespoonful of sugar. Heat these slowly in a porcelain or granitized iron kettle until it boils, then let it simmer for five minutes. Strain it through a cloth, let cool, and sweeten again to taste. It can be diluted with water.

If strawberries, raspberries, black raspberries, or blackberries are used, prepare them in the same manner, excepting that for each quart of berries a pint of water with a tablespoonful of sugar should be used.

For Grape Juice (see page 45).

Allow one pint of water to three pints of fruit (picked from the stems). Let it simmer slowly for five minutes, then strain it through flannel or cheese cloth. It is drunk cold without sweetening, although there is no law against adding a little sugar, if preferred.

Apple-water.

(The same for any of the fruits, viz.: pears, peaches, plums, French prunes, figs, raisins, rhubarb, etc.)

Boil a large, juicy apple (pared, cored, and cut into pieces) in a pint of water in a close-covered saucepan, until the apple is stewed into a pulp. Strain the liquor, pressing all the juice from the pulp. Sweeten to taste. Sometimes these fruit-waters are made with rice or barley water. To be drunk cold.

Lemonade.

Rub loaf sugar over the yellow rind of the lemon to extract the oil; add to the lemon juice (without seeds), the sugar to taste. One lemon will make two glassfuls of lemonade, the remainder of the ingredients being water and a little ice chopped fine. Lemonade should not be too strong of lemon. Sugar syrup (page 92) is always best for sweetening drinks.

Professionals serve a couple of strawberries on top, also a couple of straws.

FLAXSEED LEMONADE.

(Demulcent drink for throat and lung troubles.)

Pour a pint of boiling water on two tablespoonfuls of whole flaxseed, cover and let it steep for three hours. When cold, add the juice of a lemon, and sweeten with sugar or sugar syrup.

MILK PUNCH.

Sweeten a glass three quarters full of fresh new milk to taste, and add one or two tablespoonfuls of brandy or whiskey. Grate a little nutmeg over the top.

A professional milk-punch maker would have two tin cups, as in cut, the top of the smaller cup fitting an inch below the top of the larger cup.

The punch is shaken vigorously up and down for two or three minutes, when it is poured into a glass with a fine froth on top.

Or the milk may be poured dexterously in a long stream from one tumbler to another to produce the froth.

EGG AND MILK PUNCH.

Stir well a heaping teaspoonful of sugar and the yolk of an egg in a goblet, then add a tablespoonful of best brandy or whiskey. Fill the glass with fresh new milk until it is three quarters full, then stir well into the mixture the white of an egg beaten to a stiff froth.

EGGNOG.

Whip well together in a bowl the yolk of an egg and a heaping teaspoonful of sugar, then stir in a tablespoonful of best brandy or whiskey. Now stir in carefully the white of the egg beaten to a stiff froth, and a half pint (one cupful) of sweet cream whipped also to a froth. The egg froth and the whipped cream should be quite ready before the other ingredients are mixed together.

TOM AND JERRY.

Beat an egg (yolk and white) with a heaping teaspoonful of sugar, until it is very light—quite a froth—then mix in one or two tablespoonfuls of rum and three fourths of a cupful of boiling water. Turn this back and forth in two hot pitchers to mix well, then pour it into a hot glass. Grate a little nutmeg over the top and serve immediately.

EGG CORDIAL. (Lady St. Clair in "Dainty Dishes.")

"One tablespoonful of cream; the white of a very fresh egg; one tablespoonful of brandy. First whip the egg nearly to a froth, then add the cream and whip both together, add the brandy by degrees and mix well. Do not let it stand after it is made. This is very nourishing, and will stay on the stomach when nothing else will. The receipt was given me by the late Professor Miller of Edinburgh."

The author would suggest the addition of a teaspoonful of sugar.

A Glass of Cream.

Of all the beverages there is nothing more wholesome for a convalescent than a glass of fresh, sweet cream. It is a hearty meal in itself with the addition of a cold roll, or a health-food cracker biscuit, and perhaps a baked apple. This is preferable to a repast with tea or coffee. A glass of cream served at a Vienna *café* is partly whipped.

Tea.

Two things are necessary to insure good tea: first the water should be at the boiling-point, actually bubbling (water simply hot or steaming not answering the purpose), and, second, that the tea should be served freshly made. Tea should never be boiled, nor left over three minutes after it is made, before drinking.

Scald out well a little Chinese earthenware teapot, then throw into it two teaspoonfuls (not heaping) of good black tea (English breakfast especially recommended). Place over the fire some clear, fresh water, and when it begins to boil well, pour two cupfuls into our little teapot. Water at the *first boiling* is much better than when boiled for some time.

Let the teapot then stand at the side of the fire (without boiling) a minute.

Now serve the teapot. Do not attempt to pour the tea into the cup and carry it some distance and expect it to be *au point*, but place the teapot on the brightest of salvers. On this have a plate and the whitest of napkins, and on this again a thin, dainty cup and saucer with a bright teaspoon at the side. The little teapot takes another corner, with a little pitcher of hot water. A little fancy dish, a leaf perhaps, contains three or four lumps of loaf sugar, and a second miniature pitcher a few spoonfuls of cream. *Connoisseurs* do not drink tea with cream or milk however. On another plate is the milk toast or whatever is decided upon for the supper.

Placing this salver on a little table by the side of the invalid's bed or chair, the invalid can see the tea poured out steaming hot, while catching its pleasant aroma. (See further remarks about tea, page 1.)

COFFEE.

I once watched a *cordon bleu* making coffee in the common coffee-pot. For several reasons I believe there is no better method of making it than his.

We will not take his proportions, for the French always have coffee too strong—at least too strong for our invalid, or any one who does not care to become one. Allow two tablespoonfuls of coffee to a pint of water. Put the coffee in the coffee-pot, and pour over it about a third of a pint of boiling, bubbling water; cover the coffee-pot and let it stand until just about to boil again, when pour in the second third; and again, when this is about to boil, pour in the remainder, letting it stand until it reaches the same point, when set it back of the range for a few moments to settle. Serve immediately.

Of course proper attention must previously have been given to the even and proper roasting of the coffee, remembering that one burned berry can quite ruin the flavor of the whole. Again, the coffee is much better when the berries have been fresh roasted. If they are not fresh roasted, place them a few minutes in the oven before grinding, and it will serve to freshen them and bring out the oil. It is a good idea when coffee is fresh-roasted and still hot to mix in a little of the white of egg. It will form a very thin coating around the berries, serving to keep them fresh. They should not be ground until ready for use. The egg then serves to clear the coffee. A mixture of two thirds Java coffee and one third of Mocha insures the best coffee.

The flavor of the coffee will be altogether different if a tablespoonful of sweet, rich cream can be served with it, instead of milk or boiled milk. If cream is out of the question, use hot boiled milk, diluting the coffee always with the hot milk instead of hot water. In fact, coffee made with milk instead of water is most excellent. Sweeten the coffee with lump sugar. The Vienna coffee is served with one or two tablespoonfuls of whipped cream on top the coffee in the cup.

CHOCOLATE.

For invalids the homœopathic preparation of chocolate called "alkathrepta" is the most wholesome, for the reason that it contains no vanilla—and vanilla is a poison for an invalid. The homœopathic books all say that it is a most unwholesome if not poisonous flavoring for any one. Indeed, vanilla is used medicinally, sometimes.

For one coffee-cupful of chocolate (half-pint cup) allow one ounce or one and a quarter tablespoonfuls of chocolate and one and a quarter cupfuls of milk. Scrape the chocolate into a tin cup and mix in by degrees the quarter cupful of cold milk; stir it carefully over the fire (taking care that it does not burn) until it is a perfectly smooth paste.

When the remaining cupful of milk is boiling, sweeten it with two lumps of loaf sugar, and stir in the chocolate paste, adding a little of the boiling milk to it at first to dilute it evenly. Let it boil a minute. Stir it into a froth with an egg-whisk, and serve immediately.

A tablespoonful of whipped cream on top of the chocolate in the cup is a pleasing addition.

SELTZER-WATER AND MILK.

An equal quantity of milk and seltzer-water mixed is considered a desirable beverage, when some nourishment and a slight aperient are required.

A friend was suffering with a light attack of pneumonia in New York, and a distinguished physician (Dr. Loomis) prescribed a glassful of half milk and half seltzer-water to be taken every four hours. The alternate two hours a half glassful of vichy was to be taken. This, with a mustard plaster and perfect rest, proved all-sufficient for a rapid recovery.

BEEF TEAS AND BROTHS.

Beef Tea.

The old mode of making beef tea by boiling meat and water several hours, or even boiling it at all, was a mistaken one. An extract was thus obtained devoid of its chief nutritive element, albumen, and containing little more than stimulating qualities.

Dr. Holland, in his admirable little work on "Diet for the Sick," says:

"The albuminoid or flesh-forming principle of meats is coagulated by hot water, and either remains in the meat or is skimmed off the extract (as scum). The water has taken up the mineral salts and the flavoring principle, but is devoid of the nutriment commonly supposed to be dissolved by it. Soups and beef tea are stimulating in their effect."

One of our army surgeons prepared a receipt which was issued for the use of the army by a Circular Order. The receipt is as follows:

Beef Extract (see next receipt).

Put a third of a pound of fresh beef, finely minced, in fourteen ounces of cold, soft water, to which four or five drops of muriatic acid and a little salt (from ten to eighteen grains) have been added.

After digesting for an hour to an hour and a quarter, strain it through a sieve, and wash the residue with five ounces of cold water, pressing it, to remove all soluble matter. The liquor will contain the whole of the soluble constituents of the meat (albumen, creatine, etc.), and it may be drunk cold or slightly warmed. The temperature should not be raised above 100° Fahr., as at the temperature of 113° Fahr. a considerable portion of the albumen, a very important constituent, will be coagulated.

LIEBIG'S RECEIPT FOR BEEF TEA
is nearly the same, viz.:—Ingredients: Half a pound of finely minced raw beef (chicken or any meat may be similarly used), one pint of pure water, four drops of muriatic acid, about one half a saltspoonful of salt. Dilute the acid and salt well in three fourths of a pint of the water, then mix well with the meat. Let it stand an hour; strain through a hair sieve, and rinse the residue with the extra quarter of a pint of water.

It may be administered in a red wineglass if the patient should become prejudiced against it on account of its red color.

Baron Liebig adds: "The liquid thus obtained contains the juice of the meat with the albumen in an uncoagulated state, and syntonine, or muscle fibrine, which has been dissolved by the agency of the acid."

It seems strange that Baron Liebig, with so much knowledge of the subject, should have made his "beef extract"* so deficient in nutrient qualities as to be condemned by many eminent physicians. Dr. Dobell says: "It is important to bear in mind that Liebig's extract of meat and other similar preparations contain very little, if any, nourishment properly so called. . . . Their principal virtues belong to the class of stimulants. . . . When mixed with water they are excellent menstrua in which to administer nutritive materials,

---

* Another preparation largely sold in market.

such as eggs, oatmeal, etc.; but without such additions they are incapable of sustaining life for any length of time. Unless these facts are borne in mind a patient may easily be starved unintentionally."

Dobell further says: "Valentine's meat juice is a most useful nutrient for the sick-room. It contains albumen in solution, and hence must not be made hot. A teaspoonful in a wineglassful of water or wine is a refreshing change from the usual list of warm foods, and is very convenient for sudden use in the sick-room."

The Valentine extract will become acid and spoiled if kept too long.

BEEF JUICE.

Choose a thick slice of fresh, juicy beef without fat. A steak cut from the round (leg) contains the most juice. Broil it for only a minute, or long enough to merely heat it throughout; cut it in many places, and press out all the juice (with the aid of a beef-juice press or a lemon-squeezer) into a warm bowl. The bowl may be placed in a basin of hot water to keep warm. If no meat-squeezer is at hand, the meat may be pressed between two hot saucers, or with a strong hand. Be careful to salt the juice very slightly. Remove the globules of fat. It may be served by the teaspoonful as ordinary beef tea, or, if solid food can be taken, the juice may be poured on some dry, fresh-made toast.

BEEF TEA FOR TRAVELLING.

Chop two pounds of fresh, juicy beef, cut from the round, very fine; place it in a bowl, with one ounce of gelatine and a pint of cold water, and let it soak for two hours, occasionally squeezing the juice from the meat-pulp with the hand. At the end of the two hours pass all the juice through a fine sieve, again squeezing all the juice possible from the meat-pulp. Season it judiciously with salt and a little pepper. Bring this juice merely to the boiling-point, and pour it into an hermetically sealed glass jar (previously heated in hot water), and seal it immediately.

When wanted for use dissolve two or three teaspoonfuls of the jelly in half a cupful of boiling water, and give it to the patient hot.

A Beef Tea for Convalescents.

Soak three quarters of a pound of small-cut pieces of fresh, bright-red, lean steak (cut from the round) in a pint of cold rainwater for an hour, squeezing the beef occasionally with the hand, then place it (beef and water) on the fire. Let it come slowly to a boil, and then let it simmer for ten minutes. Pour off the tea and remove the fat; salt it slightly, and, if allowable, add the slightest bit of red pepper; add also a spoonful of fresh and well-cooked rice or barley, or dried and toasted dice of bread, or wafer crackers, or a poached egg. Serve while still fresh-made and hot.

To Make the Bread Dice, or Croutons.

Cut stale bread into dice about half an inch square, and put them in the open oven, or some place where they will become thoroughly dry; then toast them over the fire, or brown them in a hot oven, to a nice yellow color on all sides; place them in a dish at one side of the range, that they may remain warm until the moment of serving. By keeping them warm they will continue crisp until.put into the beef-tea. These bread dice are nice in any soup. Bread dice for soups are generally fried to a light brown in a little butter, but these would not be recommended for an invalid.

Or, slices of bread may be cut with little fancy

shaped cutters into pretty figures before drying and toasting, and then they may be dignified with the name of *croutons*. For convalescents it would not be amiss to butter the dice or *croutons* slightly on one side.

CHICKEN BROTH.

Cut up half a chicken (one and a half pounds) in rather small pieces, and break the bones. Do not wash it if you would save the whole juice. Put it in the cleanest of saucepans with three pints of clear cold water and a tablespoonful of rice. Bring it slowly to a boil and let it simmer for two hours, closely covered. Half an hour before it is done throw in a little sprig of parsley. When done, pass the broth through a sieve into a hot bowl, pressing the rice through with a spoon. Let it stand a moment, and then skim off the fat. Salt it with care, also add a few specks of red pepper. I hardly dare mention the red pepper, as the broth is good enough without it, and, if any is used, a cook is sure to put in too much. Or, instead of rice, granulated barley or wheat may be used for a thickening.

The broth may be served with some dainty crackers, or wafers (page 122), on a separate dish, to be broken into the broth when served; or, for a change, the rice may be boiled separately and a tablespoonful of the whole grains added after the broth is in the bowl.

MUTTON BROTH.

Cut up two pounds of the scrag end of the neck of mutton and place it in the soup-kettle with two quarts of cold water. Bring it slowly to a boil, and then place it on the range to simmer for two hours. Pass it through the sieve; season it carefully with salt and the slightest quantity of pepper (red pepper is always preferable if used carefully). If wanted immediately, skim off the fat. It is better to set it away, allowing the fat to harden on top, when it can be easily removed. When wanted to serve, heat it to the boiling-point; pour just enough in a thin soup-bowl and add a tablespoonful of fresh, well-cooked rice.

The pearled barley of the Health-food Company is a valuable thickening for mutton broth. Half to three quarters of an hour before the broth is done a tablespoonful (two ounces) of the barley may be added to the soup. When the soup is strained the barley grains may carefully be taken out with a spoon and returned afterwards, or the barley may be cooked in other water. For a change, bread dice, or *croutons*, as explained on page 103, may be added to the broth.

CLEAR BEEF BROTH WITH TAPIOCA OR SAGO.

Take four pounds of lean beef and bone (two pounds each); cut up the meat and break the bone; cover it with three quarts of clear, cold water; bring it slowly to a boil, and let it simmer for four hours. The last hour add a sprig of parsley, two or three slices of onion (previously browned in a platter with a little butter), and a slice of carrot. When done, pour the broth through the sieve. There should be about a pint and a half of broth remaining. Remove the particles of fat. Return this strained broth to a perfectly clean kettle; add the white of an egg (beaten to a thin froth) and stir it well into the broth for the purpose of clearing it; bring it all to a good boil, when place the kettle one side for a few minutes. Pass the broth through the jelly bag. If the first dripping is not quite clear, return it to the bag. Season the broth carefully with salt and red pepper, remembering that it only takes the slightest quantity of the latter; add also a tablespoonful of either tapioca or sago, prepared as follows: Soak two teaspoonfuls of sago or tapioca an hour in clear, cold water, then pour off the water and stir it into a pint of boiling water. Let it boil slowly for half an hour, pour off the water and let it steam a moment, and then it is ready to be added to the broth.

Or, the broth can be made one day and, when strained, set aside until the next day. The hardened fat at the top and the settlings at the bottom of the jelly can be easily removed. This broth will be tolerably clear, though not so much so as when cleared with the white of an egg. Sometimes a slice of lemon (without seeds) is thrown in the soup bowl just as the broth is about to be served.

Beef Broth with a Poached Egg.

Make the broth as in the preceding receipt, and, instead of tapioca, add to the bowl when ready to serve a well-trimmed and carefully poached egg.

GRUELS.

I wish to call special attention to the barley gruel made of Robinson's patent barley flour (page 29), as explained in the following receipt. The gruel is delicate and delicious in flavor and is most invaluable for nearly all conditions of sickness. I have tried the same receipt with several American preparations of barley which were good, but not to be compared with the one first mentioned. It would be well for our own manufacturers not to be outdone in supplying an article so generally useful.

Barley Gruel.

Wet gradually (stirring it smooth) half an ounce, or one tablespoonful, of Robinson's patent barley flour, with half a gill of cold water; stir well into it one gill of boiling water; add a small pinch of salt. Let it cook over the fire for five minutes, stirring it slowly part of the time, then add half a gill of hot milk. Let it again come just to a boil, then take it off the fire, stir in a teaspoonful of sugar, and it is ready to serve. Gruels are always better when served quite fresh-made and hot.

This receipt will make a coffee-cupful of gruel. One gill contains nine tablespoonfuls of liquid. Or, for a change, a most delicious blanc-mange is made by adding to the preceding receipt, when just cooked, the well-beaten whites of two eggs; stir them in smoothly and let the mixture remain a minute over the fire (stirring it meanwhile) to set the egg, though not allowing it to boil. This can be moulded and eaten cold with a little cream poured over, yet it is better when served hot.

Graham-flour Gruel.

Ingredients: Two tablespoonfuls (one and a quarter ounces), of Graham flour, or, what is much better, the granulated wheat of the "Health-food Company;" one pint and a half of water; a saltspoonful of salt (not heaping). Mix the flour with a quarter of a pint of cold water, pouring in only two or three tablespoonfuls at first, and rubbing it well to keep from lumping, then gradually adding the rest; mix in also the salt; stir in the extra one pint and a quarter of water, when it (the water) is boiling.

Boil it slowly for an hour, or until reduced one half.

Oatmeal Gruel (No. 1).

Ingredients: One heaping tablespoonful (one ounce), of oatmeal; one pint and two tablespoonfuls of water; half a saltspoonful of salt.

Rub the oatmeal smooth with two tablespoonfuls of cold water. Add the salt to the pint of water in the saucepan, and, when it *boils*, stir in the oatmeal paste. Let it boil slowly for half an hour with the saucepan partly covered.

If this gruel be made for an infant it should be passed through a sieve.

The gruel above described is that which is most frequently used. A stronger diet is made by adding one or two tablespoonfuls of cream as soon as the gruel is cooked. The gruel when cooked will be reduced to half a pint.

When a nourishing and stimulating diet is required, the gruel can be made into what is called an "oatmeal caudle" (see below).

OATMEAL GRUEL (No. 2).—(Used in feverish conditions.)

This preparation of oatmeal is given by the United States Dispensatory as follows: "Put one ounce, or a heaping tablespoonful of oatmeal, rubbed smooth in a little water, into three pints of boiling water, and boil it until reduced to two pints; then strain it, and let it cool and settle. When it is quite cold, pour the clear gruel from the sediment, add the juice of a lemon, and sugar to taste. If it is desired to have it warm, heat it before adding the lemon juice.

OATMEAL CAUDLE.

Take the half-pint of simple gruel (as was described in oatmeal gruel No. 1, and as soon as it has slightly cooled stir in a teaspoonful of sugar and the beaten yolk of an egg; return the gruel to the fire for half a minute to cook and set the egg, stirring and not allowing it to boil. Take it again from the fire and add a tablespoonful of good brandy, Jamaica rum, port, or sherry wine.

FLOUR GRUEL, OR THICKENED MILK (No. 1).

Ingredients: One heaping tablespoonful of flour, (one ounce); one pint and three tablespoonfuls of milk; salt. Rub the flour smooth with three tablespoonfuls of cold milk, then stir it into a pint of boiling milk; add half a saltspoonful of salt, and let it simmer for five minutes. It may be flavored and sweetened by adding, when cooked, a teaspoonful of sugar and a grating of nutmeg, or a dozen raisins may be boiled in the milk, and either taken out afterwards or left in for appearance sake, though they are not to be eaten.

Flour Gruel (No. 2).

When the flour gruel No. 1 is just done take it from the fire, let it cool half a minute, then stir in the yolk of an egg, beaten well with two teaspoonfuls of sugar; return it to the fire (without allowing it to boil), and stir it until quite hot again (a half-minute), then mix in smoothly the white of the egg beaten to a stiff froth. This gruel is very nice, for a change, with the beaten white of the egg added without the yolk.

Flour Gruel, of Prepared Flour (No. 3).

To prepare the flour, knead any quantity of flour with water into a ball, and tie the whole firmly in a linen cloth; put it into an iron saucepan and cover it with boiling water. Let it boil slowly (replenishing with boiling water when necessary) for twelve hours. Place it before the fire to dry, and afterwards, when removing the cloth, separate a thick skin or rind which has formed, and again dry the ball.

Receipt: Bring a pint of milk with half a saltspoonful of salt to a boil, and then stir in one tablespoonful (one ounce) of the prepared flour, previously rubbed smooth with three tablespoonfuls of cold milk; cook about three minutes.

An excellent diet for summer complaint.

Rice Gruel.

Ingredients: One well-filled tablespoonful (one ounce) of ground rice; one pint and three tablespoonfuls of milk or water; a pinch of salt. Mix and cook it the same as simple oatmeal gruel, excepting that you boil the rice gruel fifteen minutes.

This gruel is principally used for bowel complaints. If the doctor prescribe port wine or brandy, this gruel can be made with a teaspoonful of sugar and a tablespoonful of the wine or liquor added.

FARINA GRUEL.

Rub a heaping tablespoonful of farina smooth with three tablespoonfuls of milk, and add it to a pint of boiling water; add also a pinch of salt. Let it boil twenty minutes, stirring occasionally. When done, add two gills of good, sweet cream. This gruel can also be changed as was described for flour gruel—with sugar and egg added.

CORNMEAL GRUEL.

Ingredients: One pint of water; a little salt; six tablespoonfuls of milk; one tablespoonful (one ounce) of cornmeal flour. Mix the cornmeal smooth by adding gradually the milk; add also the salt, and stir it into a pint of boiling water. After it begins to boil let it simmer (uncovered) for forty minutes.

PANADA.

Sprinkle a little salt or sugar between two large Boston soda or Graham crackers, or hard pilot biscuit; put them into a bowl; pour over just enough boiling water to soak them well; put the bowl into a vessel of boiling water, and let it remain fifteen or twenty minutes, until the crackers are quite clear and like a jelly, but not broken. Then lift them carefully, without breaking, into a hot saucer. Sprinkle on more sugar or salt if desired; a few spoonfuls of sweet, thick cream poured over is a good addition for a change. Never make more than enough for a patient at one time, as it is very palatable when freshly made, and quite insipid if served cold.

Toasted bread cut into thin, even slices may be served in the same way. This is also a good baby-diet for a child over seven or eight months old.

A panada gruel may be made by adding to a cupful of boiling water, in a saucepan, a half-cupful of stale bread crumbs (without the crust) and a pinch of salt. Let it simmer five or ten minutes, or until it is, when stirred, of the consistency of gruel. It can be sweetened or not. A tablespoonful of split raisins, boiled with the gruel, makes a pleasant flavor. Sprinkle sugar over the top when served. The raisins should not be eaten. If panada is made of the new-process flour, it is as nourishing as any of the gruels.

BREADS
AND OTHER GRAIN PREPARATIONS.

Bread.

It is very important to have wholesome, sweet, and well-made bread, especially for an invalid. The new-process flours (see page 29) are indispensable for making the most nutritious white bread.

As for yeast, the Fleischman's yeast insures always sweet bread. Receipts come with these yeast cakes. The brewers' yeast is most excellent. A gill of this yeast to three and a half pounds of flour is the proper proportion. In the country the home-made yeast is generally used. Yet this is unreliable unless made by an expert.

I will give one receipt for bread which can be made of a dry yeast, that can be obtained in the country. It is made of the "National Yeast," manufactured at Seneca Falls, N. Y. "On the Raquette," where moist yeast never has ventured, and we hope never will, we had the very best and most wholesome of bread made of this yeast. The yeast packages are dated, so that only those quite fresh need be purchased.

If baking-powder is used for any of the small bread receipts, the Horsford's baking-powder is considered quite reliable and hygienic.

Hathorn's Adirondack Bread.

This bread can be made in one day, or the sponge may be made at night and finished in the morning. If made in one day the sponge must be given a warmer temperature, which will cause it to rise more quickly.

To be quite explicit, when setting the sponge at night for four loaves of bread, place in a large bread-pan three heaping quarts of sifted flour and a teaspoonful of salt. Into this mix one and a quarter cakes of National Yeast (see preceding article) which has been previously soaked for a few minutes, softened and mixed in two quarts of lukewarm water, or water at about the temperature of 96° Fahr. Mix this all well together with a spoon, and it will make a sponge a little thicker than is necessary for pancake batter. Cover this with a clean cloth, several times folded, and set it to rise over-night, selecting a situation where as even a temperature as possible, of about 70° Fahr., can be obtained. In the morning, at about seven or eight o'clock, the sponge will be found to be in bubbles, not increased, however, more than a third of its original size. At this time add enough more flour to make it a dough solid enough to handle, though not too stiff. Knead it for about ten minutes, not more, as the grain of bread does not want to be too fine. Then cover it again in the pan and set it at the side of the fire (temp. about 90°) until it has increased about double or more in size. This will require about five hours. Now separate it into loaves, knead them separately two or three minutes, and place them in the baking-pans. Cover and set them to rise for the third time in the same warm place. This will require about an hour longer, when they are ready to bake in a rather quick oven.

If it be desirable to make the bread in one day, the

sponge can be set early in the morning, say seven o'clock, and, placing it in a temperature of about 90°, it will be ready for the second handling in about five or six hours. At one o'clock it can be kneaded as before described, at five o'clock it can be made into loaves, and at six or six and a half o'clock it will be ready for baking.

A good GRAHAM BREAD can be made by preparing the sponge with white flour and mixing afterwards with the Graham flour.

GRAHAM BREAD (Quogue Receipt).

Ingredients: one cupful light bread sponge (in the morning); one cupful lukewarm water; one large tablespoonful molasses; one large spoonful of lard or nice drippings; a small half teaspoonful soda; Graham flour; a little salt.

Dissolve the soda in the water and pour it and the molasses, lard (soft), and salt into the sponge. Mix it together, then stir in as much Graham flour as you conveniently can with a spoon, making a stiff batter. Put immediately into a rectangular pan (buttered) about ten inches long, six inches wide, and four inches high. Set it in a warm place, and when well raised (or when this sized pan is even full) bake it immediately for an hour.

GRAHAM BREAD (Health-food Co.).

Ingredients: one cupful bread sponge; one half cupful warm water; two cupfuls Graham flour, or as the Health-food Company calls it, granulated wheat; one cupful cornmeal, or, without the Indian meal, three cupfuls of granulated wheat; lard the size of an egg; one half teaspoonful salt; one tablespoonful sugar. The ingredients are mixed together as directed in the preceding receipt.

Boston Brown Bread.

Ingredients: two cupfuls (one pint) milk; two cupfuls cornmeal; one cupful rye meal, or, if more convenient, Graham flour; a scant half-cupful New Orleans molasses; one scant teaspoonful soda; one teaspoonful salt; steam four hours; bake twenty minutes.

Mix the cornmeal, rye flour, and salt well together; dissolve the soda evenly first with a little of the milk, then with the whole pint. Make a little well in the flour, in which pour in the molasses, then the mixed milk and soda. Stir all well together free from lumps, and pour it quickly into a double kettle (see page 85), buttered, in which the water is already boiling. Boil it four hours, never allowing the water to stop boiling; then take out the bread and bake it for twenty minutes in the oven.

If no double kettle be at hand, pour the bread paste into a long tin pail, which cover, and set in an iron pot of boiling water, the water reaching about three fourths to the top of the pail. Cover also the iron pot, confining the steam as much as possible. As the water boils down replenish it with boiling water.

A SLICE OF BOSTON BROWN BREAD covered with cream makes a good breakfast for an invalid. A little sugar may or may not be sprinkled over.

Toast.

Cooks generally show great carelessness and ignorance in making toast. The bread slices are generally cut too thick, the crust is not taken off, and in the hurry of preparation the slices are unevenly colored, and the centre is often a mass of hot dough. Instead of a most digestible article of diet, as it should be if properly made, it becomes the most unwholesome of breads.

The slices should be cut quite thin and even, the shapes made regular by cutting off the crust and uneven sides. The scraps of bread left may be dried and saved in a can for bread-crumbing, *i. e.*, they are not to be wasted. The slices can be placed on a tin platter and dried for a little time in the open oven, or at the top of the range, when they will toast very quickly. The operation is not so quick without this drying process, for then the slices must be placed in the toaster and simply turned from one side to the other without coloring until the bread is thoroughly dried through, then it should receive a deep yellow color quite even and artistic. If allowed to color at first it will be difficult to dry the interior.

If the toast is to be served dry it should be served immediately on a warm plate; indeed, the bread should not be toasted until the person for whom it is intended is ready to eat it. If the toast is made to serve with a poached egg, a bird, or a vegetable, a little boiling water should be poured in the bottom of the plate to partly soften the toast. It should be buttered, and salted slightly also, as soon as cooked. A prettier way of serving toast is in the form of

SIPPETS.

Cut thin slices of bread into parallelogram strips; toast them carefully and evenly, without breaking, until they are crisp and golden. Serve them on a hot plate as soon as they come from the fire, arranged as in cut, and slightly buttered if there be no objection.

Bread sippets are sometimes served to an invalid with the juice from roast beef or mutton poured over. For this the bread slices need not always be toasted.

WATER TOAST.

Have an artistic piece of toast made as described in the article on toast, and, while still hot, spread a little butter evenly over the top, also a slight sprinkling of salt; pour over three fourths of a cupful of boiling water. Cover the dish with a saucer, and place it in the oven for a few minutes to soak up the water, then serve immediately.

CREAM TOAST (very good).

Toast the slice of bread as before explained; place it on a hot plate; pour over boiling water, which drain off again in a few moments, allowing the bread to become partly soft; spread over a little butter and sprinkle over a little salt, then pour over three or four tablespoonfuls of fresh, sweet cream. Let it remain in the hot oven two or three minutes to swell.

MOCK-CREAM TOAST.

Read over the article on "toast," and while two slices of bread are drying in the oven make the sauce as follows: Put in a little saucepan a cupful (one half pint) of milk; when it begins to boil stir in two even teaspoonfuls of flour, rubbed smooth with a tablespoonful of cold milk, also a pinch of salt; let it boil a minute, allowing the flour to cook thoroughly; now take it from the fire, add a piece of butter the size of a hickory-nut, and stir in the white of an egg beaten to a stiff froth; return the saucepan to the fire for a moment to set the egg, without allowing the sauce to boil. Place the saucepan at the back of the range, while you carefully toast the two slices of bread; dip them, when toasted, a moment in boiling water, then sprinkle over a little salt and the thinnest layer of butter; pour over the sauce and serve immediately.

Milk Toast.

Prepare the toast as described for "water toast," only, instead of water, pour over milk prepared as follows: Bring a cupful of milk to a boil, then stir in an even teaspoonful of flour, rubbed smooth, with a tablespoonful of cold milk; add also a pinch of salt. Let it boil a minute to cook the flour thoroughly, then take it from the fire, stir in butter the size of a hickory-nut; pour it over the toast placed in a hot dish, set it in the oven for two or three minutes to soak, then serve immediately.

Pulled Bread.

Break off irregular pieces of fresh bread about the size of an egg, and bake them in a slow oven until quite dry and slightly colored.

Pieces of stale bread or cold biscuits split in two can be made as good as new by dipping them quickly in cold water and baking them in a hot oven until the surface is crisp and the interior is well heated through.

Zwieback.

The German zwieback, which can be obtained of the bakers, is an excellent breakfast bread, to serve with a hot beverage. It is composed merely of slices of rusk dried in a very slow oven to a delicate orange color. Vienna bread slices are prepared in the same way.

The zwieback is subjected for a long time to a slow, even heat, which can be best obtained in a brick-oven.

Coffee Cake.

Ingredients: Two cupfuls of bread sponge; one egg; one half cupful of sugar; lard, the size of a hickory-nut; one cupful of warm water.

Mix these ingredients together and make a dough not quite as stiff as for bread. Let it rise well (about two hours or more); roll it out about an inch thick. It will spread over a large, square platter. Let it rise again until quite light (half an hour or more). Before placing in the oven, spread over the top one egg (both white and yolk) beaten with a teaspoonful of sugar, and again sprinkle over this about a teaspoonful of coarse, granulated sugar.

Dixie Biscuits.

This delicious biscuit I have dared to recommend for convalescents for a change of bread, as it is to be eaten cold. Like the Vienna bread, made with the same yeast, they are better quite fresh-baked, or as soon as cold.

Ingredients: three pints of sifted flour (one and a quarter pounds); one and a half coffee-cupfuls of milk (three quarters of a pint); lard, size of an egg; one egg; one third of a cake of Fleischman's compressed yeast; one teaspoonful of salt; a tablespoon even full of sugar.

The measure of milk is a pint after the lard is added. Put this mixture (the milk and lard) over the fire, and just as it comes to a boil take it off and let it get lukewarm; in the meantime put the yeast cake to dissolve in a couple of tablespoonfuls of milk, and so soon as the yeast becomes soft, rub it smooth and add it to the milk and lard when the latter are lukewarm (not before). Mix the salt and flour well together; make a well in the middle, pour in the egg, well beaten, with the sugar, then the milk, lard, and yeast. Stir all well together with a spoon, place it in a moderately warm place at the side of the range. When it has risen light (about an hour, or possibly a little longer), knead it, without adding more flour, about fifteen or twenty minutes, always stretching out the dough towards you, doubling it, and kneading on top (to form a proper grain). Cover and set it away until it has risen quite light again (about three or four hours). Then roll it out a good half-inch thick; cut it neatly with a cutter about two and three fourths inches in diameter; roll the smaller cuts left, to about half the thickness of the other, and cut it with a second cutter two inches in diameter (a kitchen pepper-box top will do). Place the small cuts on top the large ones in a platter, and do not place the large ones too near each other. When all are arranged set them away to rise for the third time (about an hour). When quite light, bake in a quick oven. If the biscuits are wanted for the invalid's six-o'clock tea, they should be begun about half-past ten o'clock in the morning.

Or the dough (without the egg and sugar if for a dinner or breakfast bread) can be made in the form  of braids, as shown in cut. This is easily done. Three rolls of even size are braided, the ends trimmed and turned under.

WAFER BISCUITS.

Rub a piece of butter the size of a large hickory-nut into a pint of sifted flour; sprinkle over a little salt. Mix this into a stiff, smooth paste, using therefor the white of an egg beaten to a froth and some warm milk. Beat the paste with a rolling-pin for half an hour or longer; the more the dough is beaten the better are the biscuits. Form the dough into little round balls about the size of a pigeon's egg, then roll each of them to the size of a saucer. They should be mere wafers in thickness. Sprinkle a little flour over the tins. Bake.

These wafers are exceedingly good to serve with an invalid's soup, or with a cup of tea, or they may be soaked in the oven with cream or milk, as described for cream toast. When made with the new-process flour or the cold-blast flour, containing the full nutrition of the wheat, these wafers, when soaked in a nutrient liquid, constitute for the invalid not only a healthful, but a sufficient meal.

Wafers of oatmeal, granulated wheat, barley-gluten, etc., or of mixtures of different grains, can be made in the same manner as the wafers described in the preceding receipt, or they may be made by simply adding a little salt and mixing with water, then beaten for twenty minutes or more.

They may be varied in design; for instance, cut into diamond shape with a knife, or with a scalloped paste jagger, or in long, narrow strips four inches long and three fourths of an inch wide, like toast sippets. However they are cut, let them be quite regular and even in shape and also baked with care.

CORN BREAD (No. 1).—(U. S. Hotel, Saratoga.)

Ingredients: Two cupfuls flour; one cupful and a

half of cornmeal; a scant half cupful of sugar; one and two thirds cupful sweet milk; two eggs; lard or butter, size of an egg (one ounce); a saltspoonful of salt; three teaspoonfuls of baking-powder.

Mix the flour, cornmeal, salt, and baking-powder well together; next beat together the sugar and eggs, and add them to the flour, etc., and at the same time the butter (melted) and the milk. Mix all well together and bake immediately.

Sour milk can be used, when a teaspoonful of soda dissolved in a quarter cupful of hot water should be substituted for the baking-powder.

Corn Bread (No. 2).

Ingredients: one pint cornmeal; one pint of sweet milk; one egg; one tablespoonful of sugar or syrup; one teaspoonful lard (melted); two teaspoonfuls baking-powder. If sour milk is used, a half teaspoonful of soda instead of the cream of tartar should be substituted.

Corn Bread (No. 3).—(Very good.)

Ingredients: one cupful and a half of milk; one cupful of fine cornmeal, sifted; two eggs; scant tablespoonful of butter; one teaspoonful of sugar; one teaspoonful of baking-powder. Pour the milk, boiling, on the sifted meal. When cold, add the butter (melted), the salt, sugar, baking-powder, the yolks of the eggs, and lastly the whites, well beaten separately. Bake half an hour in a hot oven.

Or the corn cake is still better as follows:—Ingredients: one pint of milk; half a pint of cornmeal (sifted); four eggs; a scant tablespoonful of butter, salt, and one teaspoonful of sugar.

This last receipt contains no baking-powder. The whites of the eggs should be well beaten to a stiff froth. The ingredients are put together exactly as described in first receipt.

Corn Rice Bread.

Ingredients: one half pint of cornmeal (one cupful); one pint of cold boiled rice; one half pint (one cupful) of milk; one egg; one half teaspoonful of salt; one tablespoonful of sugar; butter, size of pigeon's egg; one teaspoonful of baking-powder. Mix the baking-powder, sugar, salt, and cornmeal well together. Pass the rice through the collander, and add it to the milk, egg, and butter (melted). Then stir in the cornmeal, etc., and put it quickly in the oven.

Hoe Cake.

Pour enough boiling water and milk mixed (say half and half—the milk causes it better to brown) on cornmeal, salted, to make it rather moist. Let it stand an hour or longer. Put two or three heaping tablespoonfuls on a hot griddle greased with lard. Smooth over the surface, making a flat cake about half an inch thick and of round shape. When browned on one side, turn, and brown it on the other. Serve very hot. A good breakfast cake with a savory crust.

Pancakes (of Flour, Granulated Wheat, Cornmeal, Bread Crumbs, Oatmeal, Rice, Gluten, etc.).

Stir two cupfuls of milk into two beaten eggs, and stir in enough of any of the flours to make a thin batter; add a little salt, and then sprinkle over and stir in well a heaping teaspoonful of baking-powder (if the milk is sweet) just before baking.

If there is any cold boiled rice, oatmeal porridge, hominy, etc., at hand, some of any or all of these improve very much the pancakes. The pancakes are also better the whites of the eggs are beaten to a stiff froth, and this and the baking-powder are added just when the cakes are to be cooked (not before).

If sour milk is used, a scant half-teaspoonful of soda dissolved in a little warm water should be stirred in the last thing, although more or less of soda is used, according to the acidity of the milk. If the griddle is quite hot and smooth, and is merely moistened with a little lard, the cakes will not be greasy nor so very unwholesome. However, I will not risk recommending them for our invalid.

CURRANT SCONE ("Hygienic Cookery").

Ingredients: two cupfuls sifted Graham flour; two cupfuls sifted white flour; one cupful and a half of thin sweet cream—part milk will do; one cupful and a half of English currants, picked, washed, and drained; two and a half teaspoonfuls of baking-powder, or two-thirds teaspoonful of soda, and one and a half teaspoonfuls of cream tartar. Stir together the Graham and white flour, add the soda (pulverized) and cream tartar (or, in its place, the baking-powder), and sift two or three times. Then stir in the currants, and wet with the cream to make a tolerably stiff dough; knead as little as possible; gather up the mass lightly till it will stick together, and roll to the thickness of half or three quarters of an inch. Prick deeply with a fork or draw shallow lines across the top, forming diamond creases. Bake.

It is very good made of Graham flour without the white flour. It is not good the day after it is baked. For an invalid it is a good cake, eaten with grape juice, etc.

HARD GRAHAM ROLLS.

This is a bread much used by the hygienists, and is called "the perfect bread" by Dr. Trall. It is much relished by those who have become accustomed to it, and who crave "no spice but hunger, no stimulant but exercise."

It is made by simply mixing cold water—the colder the better—into good Graham flour until it becomes a moderately stiff dough, and kneading or pounding it, like the Southern beaten biscuit, for twenty minutes or more, until it becomes smooth and elastic to the touch, and brittle if pulled. If the dough is too stiff the biscuit will be dry and hard, and if too soft it will be wet and clammy. It will require, perhaps, two thirds of a pint of water to mix a quart of flour, although the quantity will vary according to the grade of flour. The best of Graham flour is made by Ferdinand Schumacher, of Akron, Ohio, and also by the Health-food Company. The dough is formed into little biscuits about three inches long and not quite three fourths of an inch wide. Make out the panful quickly, setting them a little apart; prick them with a fork, and bake in a rather quick oven. When done they should not yield to the pressure of the finger. They may be made into the form of stems of the shape of lady's fingers.

These rolls are better baked fresh, although if any are left from the day before they are most excellent when warmed over, as follows: Break each roll into two or three pieces (do not cut them); drop them into cold water, and when soaked place them on a bread-pan in a brisk oven which will crisp without shrivelling them. As soon as stiff and lightly crisped they are done.

CRACKED WHEAT.

The receipt here given is undoubtedly the very best one for cooking cracked wheat, or the whole grain of wheat. It is one of the most important receipts in the book, for the invalid, or, indeed, for any one. It supplies a dish very palatable, and although light and wholesome enough for the most delicate stomach, it is as hearty in the richness and fulness of its nourishing qualities as a full meal of meat and vegetables. It was a very favorite dish at the Vienna bakery in St. Louis, and many went there for the purpose of taking it for luncheon. Several spoke to me about it, extolling its merits; but as cracked wheat had been an unpalatable dish to me as compared with oatmeal, I was slow to try it. As soon as tried, however, served as never seen be-

CRACKED WHEAT.

fore, viz., cooked with milk, each grain lying separate in a cream jelly, served cold in a moulded shape and with a little pitcher of cream, and a bright silver urn containing pulverized sugar accompanying it, I became an immediate convert, and it has been a very frequent dish at home ever since.

It is well to ask for the cracked wheat, double milled, at the grocer's or miller's, if you would avoid the silicious fibre which encircles the grain, and which is sometimes unwholesome for those with delicate stomachs. The preparation I prefer is the whole grain, as sold by the Health-food Company of New York, which is quite free from the woody skin.

Receipt. — The ingredients are, one half cupful of cracked wheat; two and a half cupfuls of water; two and a half cupfuls of milk; one half teaspoonful salt.

Salt the water, and when it comes to a boil add the grits and let it simmer, without cover, on top of the range for an hour. The water will then be almost evaporated; then add the milk (hot) and let it cook an hour longer. Stir it occasionally to keep the wheat from attaching at the bottom, and also to mingle evenly the grains with the liquid. More stirring than this is objectionable. A copper or porcelain saucepan or earthen crock is preferable for cooking this dish, on account of less danger from burning. The wheat cooked in a double kettle will not be as good, the steam puffing through the grains giving better flavor. There is no danger of burning if not cooked too fast. The milk used should be perfectly fresh and sweet, or the mixture will curdle.

When done, stir it carefully, as it will be thin and the grains will be liable to sink, and pour it into cups (previously wet with cold water) about three fourths full. Set them one side to become cold and solid. Do not remove the wheat from the moulds until ready to serve. Serve with cream or milk and pulverized sugar.

OATMEAL PORRIDGE.

It seems very simple to make oatmeal porridge, yet it is a very different dish made by different cooks. The ingredients are: one even cupful (one half pint) of oatmeal to one quart of boiling water, and one teaspoonful of salt. Boil forty-five minutes.

The water should be salted and boiling when the meal is sprinkled in with one hand while it is lightly stirred with the other. When all mixed it should boil slowly, uncovered or partly uncovered, without afterwards being stirred more than is necessary to keep it from adhering to the bottom, and to mingle the grains two or three times that they may all be evenly cooked. If much stirred the porridge will be starchy or waxy and poor in flavor. The puffing of the steam through the grains without much stirring swells each one separately, and when done the porridge is light and palatable.

EARTHEN CROCK.

Professional cooks insist upon having copper saucepans for cooking the grains, for the good reason that there is but little danger of burning in them. A common earthen crock placed on top the range answers the purpose very well. Care must be taken that a cold crock should not suddenly be placed on a very hot surface. Pour hot water in the crock before placing it on the range, and there will be little danger of breaking. This manner of cooking is applicable to all the grains.

CORNMEAL MUSH.

This may be made by stirring, say, a pint of cornmeal into three pints of salted boiling water, and cooking it a good half-hour. Or, stir a pint of cornmeal, mixed with a pint of milk and a teaspoonful of salt, into a quart of boiling water, and let it boil half an hour, stirring often.

RECEIPTS FOR GLUTEN.

I FIND it a little difficult to provide very palatable dishes out of gluten, without starch. Added to rice, farina, and other starch grains, which are prohibited in some diseases, it is very palatable made into pancakes, or any of the puddings made of other grains. For thickening sauces, soups, or gravies, it is very satisfactory. Gluten used instead of bread-crumbs for egg and crumbing fish slices or fillets, oysters, sweetbreads, etc. (for frying), is also a success.

GLUTEN BREAD.

Ingredients: one pint of milk; one pint of warm water; butter or lard size of a walnut; one half cake of any fresh, dry hop yeast, or one fifth of a two-cent cake of compressed yeast, rubbed smooth with a little water; one egg, well beaten; a little salt.

Mix the milk, water, egg, yeast, and lard (melted), and stir in the gluten until a soft batter is formed. After it has risen (in some warm place) mix in gluten enough to form a soft dough (like biscuits), and knead well. Form into loaves, and, when risen a second time, bake. Gluten bread requires less yeast than ordinary bread, and less time in rising.

GLUTEN MUSH.

Place one and a half cupfuls of water on the fire to boil. Stir smoothly either a cupful of cold milk or water into a cupful of gluten, and a half-teaspoonful of salt. When the water boils, pour in the mixture gradually and let it cook twenty minutes.

FRIED MUSH.

Slices of cold gluten mush fried or *sautéd* in a little hot lard.

GLUTEN MUFFINS.

Ingredients: one cupful and a half of gluten; one cupful of milk; one egg; one-fourth teaspoonful of salt; one teaspoonful of baking-powder.

Heat the gem pans before buttering, pour in the batter, and bake fifteen minutes in a quick oven.

This quantity will make eight gems, or just fill one of the ordinary iron gem pans. Or, the flavor is better to add rice as follows:

GLUTEN AND RICE MUFFINS (not for diabetics).

Ingredients: one cupful of gluten; one cupful and a half of cold, boiled rice; one cupful of milk; one egg; one half teaspoonful of salt; butter size of hickory-nut; two teaspoonfuls of baking-powder.

Mix the baking-powder, salt, and gluten well together. Pass the rice through a colander, and stir into it the milk, egg, and butter (melted); next add the gluten mixture, and put it quickly into the oven. Or, instead of rice, the same quantity of cold, boiled pearled barley, or oatmeal may be substituted; or three fourths of a cupful of cornmeal and one cupful of gluten, with the other ingredients in the preceding receipt, make good breakfast muffins.

A GLUTEN PUDDING OR GRUEL.

Ingredients: one cupful of water; two tablespoonfuls of gluten, rubbed smooth in four tablespoonfuls of cold water; the white of one egg; salt.

When the cupful or half-pint of water is salted and boiling, mix in the gluten paste and let it cook ten minutes; stir in then the white of an egg beaten to a stiff froth. Let it remain a half-minute (while stirring it) to set the egg. To be eaten hot, and fresh made. Or, instead of four tablespoonfuls of cold water for making the gluten paste, let it be four tablespoonfuls of cream, and the pudding can be sweetened with a scant tablespoonful of glycerine.

Gluten Pudding.

Soak two slices of gluten bread in a little milk in which an egg, a tablespoonful of glycerine, and a sprinkling of nutmeg have been mixed. Do not let the bread get too soft to handle. Fry the slices on a griddle in either a little hot lard or butter.

Gluten Cream Wafers.

Stir gluten (crude or purified) and a little salt into sweet cream, until the dough is thick enough to roll out to the thickness of pasteboard. Beat the dough with a potato-masher for fifteen minutes or more, roll out, cut into forms, and bake.

Gluten Cheese Cakes.

Add to a cupful of gluten three tablespoonfuls of grated cheese, two tablespoonfuls of cream, the yolks of two eggs, a saltspoonful of salt, and a little nutmeg. Roll thin and bake like cookies.

Gluten Soufflé.

To a half cupful of gluten add two tablespoonfuls of grated cheese, the beaten yolk of an egg, half a saltspoonful of salt, and three tablespoonfuls of cream. Mix this evenly together, forming a soft paste a little thicker than for pancakes. The last thing stir in the whites of the eggs beaten to a stiff froth. Bake in patty pans, or paper cases, and serve as soon as baked. It is a very rich dish, too rich for much to be eaten at one time.

VEGETABLES.

A Baked Potato.

A Potato baked, when properly prepared, is probably the most digestible form in which it can be served. The excellence of a baked potato depends much upon its being served *immediately* when cooked to a turn. A moment underdone and it is indigestible and worthless; a moment overdone and it has begun to dry. It requires about an hour to bake a large potato in a hot oven. When served and mashed, the addition of some cream and a little salt is most excellent.

To Boil Potatoes.

Choose those of equal size. Take off a very thin peeling, as the best of the potato lies nearest the skin. Put them into enough well-salted cold water to cover them; let them boil till thoroughly done, and do not let them remain a moment longer. Drain off the water, cover them closely, and set the vessel at the side of the fire, to allow them to steam for several minutes. A point is made in keeping the potatoes covered while steaming, for the purpose of retaining heat enough to draw out the moisture. The escaping moisture, though covered, will not return to the potatoes. Sprinkle over some salt as soon as they are fully steamed. It requires about thirty-five minutes to boil medium-sized potatoes.

A copper saucepan, or an iron pot retaining an even heat, should be used for boiling potatoes—never a tin saucepan.

POTATOES (*à la crême*, very good).

Cut cold, boiled potatoes into little square bits or dice, say a third of an inch square; mix them with enough white sauce to moisten them, made as follows:

Place a tablespoonful of butter in a small saucepan, and when it bubbles throw in a tablespoonful of flour; cook it a minute without coloring, then add a pint of milk or half milk and half cream, season with a level teaspoonful of salt, a pinch of pepper, and a little nutmeg. This will make a pint of cream sauce, and will be sufficient for a quart of potatoes.

Place a little butter or drippings in a frying-pan (or *sauté* pan), and, when hot, put in the moistened potatoes, color them on one side, loosen them from the pan with a pancake-turner, turn them like an omelet on a platter, and serve.

POTATOES (*à la crême, au gratin*).

Delmonico serves potatoes as prepared in the preceding receipt, and instead of *sautéing* (or frying) them, they are placed in a basin or pudding-dish sprinkled over with cracker-dust and a little grated cheese, and then they are colored in the oven. It is perhaps better after they are thoroughly heated in the oven to color them with a salamander or hot shovel, leaving no chance for the potatoes to become dry by too long a process of heating.

A PRETTY SPINACH DISH.

In picking over the spinach separate the thick stalks from the leaves. A bright green color is given to it by throwing it into plenty of well-salted water, when it is boiling very fast. It should be taken out the moment it is soft, for allowing it to remain too long would impair its color. Drain it well, and do as you please about putting it through a colander. Just before serving reheat it on the top of the range, adding a little butter, pepper, and salt. Serve enough for one person on a little square piece of toast, flatten the top, and decorate it with some finely chopped hard-  boiled egg, the yolk thickly sprinkled in the centre and a circle of white around. This will resemble a sunflower.

BEETS OR CARROTS (*à la crême*).

Boiled beets or carrots, sliced, are mixed in cream sauce as described for potatoes *à la crême*, excepting, in the place of nutmeg, a tablespoonful of finely minced parsley is added. The appearance of the vegetables is improved by cutting them with fancy vegetable cutters. There must not be too much sauce, only a soft coating around each slice of beet or carrot.

CAULIFLOWER (*à la crême*).

The boiled cauliflower, cut into flowerets, is mixed with cream sauce as described for potatoes *à la crême*, and, when placed in a dish for serving, the top is sprinkled over with rather coarse bread crumbs, which have been colored (*sautéd*) in a little butter.

Sometimes the top is sprinkled with sifted cracker crumbs and grated cheese, and is then colored with a red-hot shovel. Served in shells or paper cases the dish is especially attractive. Sometimes the sauce is finished by stirring in the beaten white of an egg just before it is taken from the fire. It makes also a good sauce for asparagus.

Stuffed Tomatoes (Chef Cuppinger).

For eight tomatoes make a stuffing as follows: Ingredients—Butter, size of an egg; half an onion, cut fine; three fourths of a cupful of either chicken livers or cold, cooked chicken or meat of any kind, chopped fine; three sprigs of parsley, chopped fine; one and a half cupfuls of bread crumbs, after they have been soaked in water and squeezed dry by wringing in a clean towel; one large tomato, cut fine; one egg; half a saltspoonful of thyme; a pinch of cayenne pepper; salt.

Place the butter in a saucepan, and when it bubbles add the minced onion. When it has colored slightly add the meat, bread crumbs, and all the other ingredients.

Fill the tomatoes (with the tops cut off and interior partly removed) with this mixture, letting it rise from a half of an inch to an inch above the tomato.

Place the stuffed tomatoes in a little baking-pan, sprinkle cracker crumbs over the tops, also a bit of butter over each one. Bake them in the oven about fifteen or twenty minutes.

It should be served with a brown sauce made as follows:

Brown Sauce.

This is made with but little trouble, although there are many kinds of brown sauces.

In a small saucepan place butter the size of a walnut, and when it bubbles throw in a tablespoonful of minced onion; when beginning to color add a tablespoonful of flour, which allow to color also. Now add one and a half or two cupfuls of stock if you have it, and, if not, water, and two or three sprigs of parsley. Let it cook a couple of minutes, season with a little pepper and salt, pass it through the gravy-strainer, and add one or two tablespoonfuls of almost any kind of wine—sherry being generally used.

STUFFED PEPPERS.

This is an especially nice dish of Chef Cuppinger's. As a course for a luncheon or dinner it may be better than for the invalid. Yet, as an appetizer, it would not be unfit sometimes for the latter. Use the green or red peppers of round shape; cut them lengthwise, and remove the interior, seeds and partitions; cover them with cold water and parboil them five minutes. Now proceed with them as for stuffed tomatoes, serving them also with the brown sauce.

Care must be taken not to have too much cayenne pepper in the stuffing for the peppers. None at all is really needed.

LITTLE DISHES.

Boiled Eggs.

Eggs are generally boiled by placing them in boiling water, and boiling them two and three quarter minutes. It is better to put the eggs in a saucepan of *cold* water, half a pint to each egg. Set it over a fire hot enough to make the water boil in three or four minutes. As soon as the water boils, remove the saucepan from the fire and let the eggs remain in the water one minute.

Poached Eggs.

This is probably the best mode of serving them for an invalid, unless served uncooked, as described in the succeeding receipt.

Poached eggs are generally wretchedly cooked by non-professional cooks. They are either thrown into rapidly boiling water and torn into pieces, or are overdone. If overdone they are indigestible. The albumen or white of the egg shrinks and becomes hard and tough if overcooked—indeed, it forms a cement when heated above a certain point.

The white of the egg, to be properly poached, should be white, but of a soft, transparent, jelly-like substance. It should be tender and delicate, evenly cooked throughout, no part being hard while another is semi-raw. To prepare it in this manner the water in which it is cooked should not reach the boiling-point.

The easiest way is to slip the egg (previously broken into a saucer) carefully into salted water which is simmering. Then immediately set the saucepan at the side of the range (to prevent the water from boiling) and let the egg remain about ten minutes.*

Let the water be about two inches high in a low saucepan. Each egg should be broken separately into a saucer and slipped very carefully into the water. When cooked just enough take out the egg with a perforated ladle (there should be nothing to trim), and slip it on a thin, buttered, and slightly salted square piece of toast which has previously been partly moistened by pouring a little boiling water in the bottom of the platter, and allowing the toast to soak it. As soon as cooked, sprinkle salt and a little pepper over the egg tops. Any substance absorbs more readily the flavor of seasoning when it is hot rather than when lukewarm or cold.

Poached eggs are very good introduced into beef broth. Delmonico serves poached eggs on toast with sorrel sprinkled over the tops. Fine water-cresses make a pretty garnish.

---

* W. Matthieu Williams in "The Chemistry of Cooking" says the perfection of egg-poaching is to keep the egg in water at the temperature of 160° for half an hour.

A Raw Egg.

This is an invaluable preparation for an invalid.

Beat well the yolk, together with a teaspoonful of sugar in a goblet; then stir in one or two teaspoonfuls of brandy, sherry, or port wine; add to this mixture the white of the egg, beaten to a stiff froth. If properly beaten it should fill a goblet to overflowing. Carefully stir altogether. If wine is not desired, flavor the egg with nutmeg. It is very palatable without flavoring at all, using only the sugar.

A Beefsteak.

For our invalid, cut out the tender part of the beef from the porterhouse or tenderloin steak. Let it be three quarters of an inch thick. Do not pound it. A well-shaped piece cut from the round or sirloin steak is not to be despised, as it contains more juice than the tenderloin. A cut from a round steak should not be as thick as a tenderloin cut, and, if tough, can be pounded a little. Have the gridiron quite hot and well greased with pork or beef suet. Put on the steak over a hot, clear fire, and cover it with a baking-pan. A wood or charcoal fire is preferable to hard coal for broiling anything. In a few moments, when the steak is colored, turn it over; watch it constantly, turning it when it gets a little brown. Do not stick a fork into it, as that will let out the juice, and do not place anything over it which can touch the top, as that will prevent the steak from swelling. Do not put on the pepper and

salt before the steak is cooked, as it is calculated to harden the fibres. If the steak is very thick, either the fire must not be too brisk or it should be turned very often. However, the quicker any article to be broiled is cooked, the better. When cooked enough (from five to ten minutes), it should be rare or pink in the centre, though not raw. Place it on a hot platter, sprinkle it with pepper and salt, and spread over some sweet, fresh butter; set the platter in the oven for a few moments to let the butter soak a little in the steak, then serve immediately. A steak is much improved by a simple addition called *à la maitre d'hotel,* as follows:

When the steak is cooked and placed on a hot platter it receives first a sprinkling of pepper and salt. Then comes a sprinkling of very finely minced parsley, then some drops of lemon juice, lastly small pieces of butter are carefully spread over. The steak is then placed in the oven for a few moments for the butter to become melted and soaked into it.

If an invalid can eat a beefsteak he can generally eat some one vegetable with it, and to make the little plump, tender morsel of beef look more tempting, garnish it with a vegetable.

If with potatoes, bake one or two equal-sized potatoes to a turn; when quite hot remove the inside, mash perfectly smooth, season with butter, or, what is better, cream and salt, and press it through a colander. It will look like vermicelli. Place it in a circle around the steak, or in banks on each side. Other vegetables, if allowed, as pease, string-beans, green corn, etc., can be served in the same manner.

A tomato sauce (page 155) is a most excellent accompaniment for a beefsteak.

A beefsteak is always more attractive garnished with parsley, or any kind of leaves, and slices of lemon.

CHOPPED BEEFSTEAK.

From Miss Juliet Corson's very valuable receipts for the sick, published in *Harper's Bazar:*

"Trim the fat from a pound of round or sirloin steak, cut the meat in inch pieces, put it into a meat chopper or mincing-machine, and chop it for five minutes; then take from the top of the meat the fine pulp which rises during the operation of chopping; continue to chop and to remove the pulp until only the fibre of the meat remains. Press the pulp into a round flat cake, and broil it over a very hot fire for about five minutes on each side; season it lightly with salt and Cayenne pepper, and a little butter, and serve it hot.*

In selecting beefsteak for invalids some persons choose the *filet*, or tenderloin, because it seems most tender; it is hardly more digestible on that account, for its looseness of fibre does not favor complete mastication; and it is less nutritious than sirloin or round steaks, because its muscular tissue is not so well nourished as that of the last-named cuts. Beef for the use of invalids should either be broiled quickly over a very hot fire, and lightly seasoned with salt and Cayenne pepper, roasted at an open fire, or baked in a very hot oven without any water in the pan; if the inside of beef is purple, it is not sufficiently cooked to be easily digested; the color of

---

* This steak is often served almost entirely uncooked. The pulp is slightly seasoned before it is formed into cakes, then merely heated through, although colored a light brown on the outside.

properly cooked beef is pinkish-red. The inner cuts are the most digestible."

BEEF SANDWICH.

Scrape very fine two or three tablespoonfuls of fresh, juicy, tender, uncooked beef; season it slightly with pepper and salt; spread it between two thin slices of slightly buttered bread, cut it neatly into little diamonds and serve.

A VENISON STEAK.

A venison steak should be cooked in the same manner as a beefsteak. A little melted currant jelly is a pleasant addition. It is sometimes made in the form of a sauce by diluting the jelly with a little water, and thickening it with a little cornstarch or flour.

A MUTTON CHOP.

A cut from the loin is best. One containing a large tenderloin could be chosen for our invalid. Let it be cut thick and leave on it plenty of the fat. Broil as described for beefsteak. Serve with mashed potatoes or other vegetables, and decorate it artistically.

BREAST OF CHICKEN.

For an invalid a chicken fricassee or a tender bit of boiled chicken is most desirable. A breast of a tender chicken, seasoned and rubbed with butter, and thrown on some burning charcoals which are not too hot, is very savory. If skilfully cooked the surface will be very little charred, and the inside will be very tender and juicy. When done, season again with butter, pepper, and salt.

Or another mode of cooking a breast of a spring chicken is to stick the leg bone into the end, giving it the form of a cutlet, rub it with butter, and broil it carefully. The second joint of a leg of a chicken contains more juice, and has more flavor than the breast.

A Fricassee of Chicken.

Cut two chickens into pieces. Reserve all the white meat and the best pieces; the remainder use to make the gravy. Put the latter pieces into a porcelain kettle with a quart of cold water, one clove, pepper, salt, a small onion, a little bunch of parsley, and a small piece of pork; let it simmer for half an hour, and then throw in the pieces for the fricassee; let them boil slowly until they are quite done, take them out then, and keep them in a hot place. Now strain the gravy, take off all the fat, and add it to a *roux* of half a cupful of flour, and a small piece of butter. Let this boil a few moments, then take it off the fire and stir in three yolks of eggs, mixed with two or three tablespoonfuls of cream, also the juice of half a lemon. Do not let it boil after the eggs are in or they will curdle. Stir it well, keeping it hot a moment; then pour it over the chicken and serve. Some of the fricassees with long and formidable names are not much more than wine or mushrooms, or both, added to this receipt.

Chicken Croquettes (Philadelphia Cooking School).

To every pint of cold cooked chicken, chopped very, very fine, allow half a pint of cream or milk, one tablespoonful of butter, two tablespoonfuls of flour, one tablespoonful each of parsley and onion, chopped also very fine, a little nutmeg, salt, and Cayenne pepper to taste.

Place the butter in a saucepan, and when it bubbles

throw in the onion, parsley, and flour, and let them cook a minute without taking color; then pour in the milk,  stirring it well with an egg whisk until the mixture is quite even and smooth. Let it boil another minute to thoroughly cook the flour, then stir in the chicken pulp and seasoning. When cool, foam into croquettes, roll in beaten egg and sifted cracker crumbs, and fry by immersion in boiling lard. The paste will be rather soft to handle, but a cook can easily manage it with a little practice. Of course, the softer the paste, the more creamy and soft will be the croquettes when cooked.

Croquettes are very good made with finely minced cold roast veal (not boiled) instead of chicken. They imitate the chicken croquettes in flavor.

Or, they can be made of cold roast beef, roast lamb, mutton, cold cooked sweetbreads, cold fish, etc., instead of chicken. In case sweetbreads are used they are cut into dice rather than minced.

Chicken croquettes are much improved when served with a sauce, either brown, white, or tomato sauce. They are sometimes served with pease, etc.

CHICKEN WITH MACARONI, OR WITH RICE.

Cut the chicken into pieces; fry or *sauté* them in a little hot drippings, or in butter the size of an egg; when nearly done put the pieces into another saucepan; add a heaping teaspoonful of flour to the hot drippings, and brown it. Mix a little cold or lukewarm water to the *roux;* when smooth add a quart or more of boiling water. Pour this over the chicken in the saucepan, add a chopped sprig of parsley, a couple of slices of onion, pepper, and salt. Let the chicken boil half or three quarters of an hour, or until it is thoroughly done; then take out the pieces of chicken. Pass the sauce through a sieve, and remove all the fat. Have ready some macaroni which has been boiled in salted water, and let it come to a boil in this sauce. Arrange the pieces of chicken tastefully on a dish, pour the macaroni and sauce over them, and serve. Or, instead of macaroni, use boiled rice, which may be managed in the same way as the macaroni.

Plain Boiled Chicken.

Throw the chicken, cut into pieces, in plenty of boiling water (enough to have some left, after the boiling is over, for sauce). Boil slowly until the chicken is very tender, if it takes all day. Thicken the gravy with flour, first rubbed smooth with a little cold water. Season with pepper and salt. A potpie addition is generally made to this dish.

Fried Spring Chicken.

The excellence of spring chickens depends as much on feeding as on cooking them. All chickens should be drawn as soon as killed, and are better if killed a day before cooking. Do not wash them. Several

hours before cooking the chicken, dismember it, and dip each piece hastily in a bowl of water; spread them on the table, sprinkle pepper and salt over all, then turn and season also the other side. Roll each piece separately, while still wet, in a plate of flour. When ready to cook have two or three spoonfuls of lard in a *sauté* pan or spider quite hot, in which fry, or, rather, *sauté*, the chickens, covering them and watching that they may not burn. The quicker they are cooked without scorching the better. When done arrange them on a hot dish, pour out the lard from the spider, leaving what will stick at the bottom. Pour in one or two cupfuls of milk, thicken it with a little flour (rubbed smooth with a little cold milk), season with pepper and salt, pass it through the gravy strainer, pour it over the chicken. Minced parsley is often added to the gravy. A circle of boiled rice or cauliflower around the chicken with the white sauce poured over both is very nice. Decorate with parsley.

Chicken Soufflé.

Chop half a pound of cold cooked chicken (freed from skin and bone) fine as possible; pound it in the chopping-bowl, or, better, in a mortar; then rub it through a sieve with the edge of a large spoon. The white meat, although it has not the flavor of the dark meat, is better suited to this purpose.

Now make a *roux* in a saucepan as follows: Place in it butter size of a pigeon's egg, and when it bubbles stir in, with an egg whisk, a dessertspoonful of flour; when evenly blended stir in three quarters of a cupful of hot water, and let it cook a few moments, stirring it smoothly together with the egg whisk; then stir in the chicken pulp, and season it palatably with salt and a little red pepper. Let the paste get entirely cold (covering it so that it will not get hard), then mix into it lightly, first the yolks of two eggs beaten to a cream, then the whites of three eggs beaten to a stiff froth. Put it immediately into little paper soufflé cases, or silver scallop shells, or into a little pudding dish. Bake about fifteen minutes in the oven, and serve it immediately when done.

A Bird

Broiled, as described for beefsteak, and served on toast, is good for an invalid who is very well, provided the bird is quite tender. It is not to be trusted for a genuine invalid.

Breast of a Prairie Chicken.

The breast of prairie chicken broiled and served on toast is most digestible if tender. If not very tender it should be parboiled before broiling. Sometimes it is boiled with a little onion and parsley added to the water, and when done the gravy is strained and freed from fat, thickened with a *roux* (flour and butter), and seasoned with some claret or sherry.

Broiled Fish.

For this purpose a white fish from the lakes, or a bass is generally used. The two sides of the fish are spread open by cutting partly through the back. It is seasoned with pepper and salt and sprinkled well with flour. The inside of the fish is first presented to the fire on a gridiron, well greased with lard or a piece of pork. As the fish can only be turned once, it must be watched carefully to avoid burning. Before turning, loosen the fish carefully from the gridiron with a knife or pancake turner. If large, place a platter close over the top, and, turning the gridiron, the fish is left in the platter, when it can be easily slid to the gridiron again, for the purpose of cooking the other side.

When cooked, serve the inside of the fish uppermost on the platter, sprinkle over pepper, salt, and butter, minced parsley, and a little lemon juice. Place it in the oven for a few moments to soak the butter, etc. Garnish with lemon slices and parsley.

Boiled Fish

Is cooked by first immersing it in cold salted water. It is generally served with a drawn-butter sauce, with an addition of chopped hard-boiled eggs, or minced parsley, etc. Sometimes the fish is cut transversely into pieces about an inch and a half long and cooked "*en matelote*," as follows: sprinkle salt on them and let them remain while you boil two or three onions (sliced) in a little water. Pour off this water when the onions are cooked, and add to them a little pepper, about a teacupful of hot water, and a teacupful of wine, if it is claret or white wine, and two or three tablespoonfuls if it is sherry or port; now add the fish; when it begins to simmer, throw in some bits of butter which have been rolled in flour. When the fish is thoroughly cooked (about fifteen minutes) serve it very hot. Stewed fish is much better cooked with wine, but is very good without it, in which case add a little parsley. Decorate the dish with fancy cuts of toasted bread.

Bass à l'Espagnole.

Cut a bass or a flounder into *filets* as follows: Lay the fish on the table, and with a thin, sharp-bladed knife cut down to the bone in the centre of the fish, following the course of the backbone, from the head to the tail. Insert the knife in the cut already made and cut towards the fin, keeping the knife pressed close against the bone, taking off the whole side piece, or *filet*. Take care not to mangle the flesh. Cut off all four of the side pieces of the fish in the same way, and lay them with the skins downwards on the table, holding the end of a *filet* with the fingers of the left hand, lay the blade of the knife flat on the table between the skin and meat, cutting from you. If the end is held firmly, the knife laid flat, the whole *filet* can be cut from the skin, without mangling it.

Broil the *filets* on an oiled gridiron, over a moderate fire, spreading a little butter, pepper, and salt over them as they are cooking. Lay them on a hot dish and pour over them a sauce made as follows: Fry the slices of a quarter of an onion, partly coloring them in a little hot butter; at the same time a teaspoonful of flour may be thrown in to receive also a little color. Pour in now a cupful of stock and a cupful of canned tomatoes, season with cayenne pepper and salt, and when it has boiled a couple of minutes, to become slightly thickened, pour it over the cooked *filets* without straining. Over the top of the dish sprinkle very finely minced parsley. Professional cooks sometimes add, also, minced mushrooms to the sauce.

SWEETBREADS.

Professional cooks generally soak sweetbreads for an hour in cold water before cooking, for the purpose of making them white. The flavor is better, however, to throw them immediately into boiling salted water, and let them cook rapidly until thoroughly done (about twenty minutes). Remove, then, the skin and little pipes, sprinkle over pepper and salt, roll them in egg, peppered and salted, and then in fine sifted cracker crumbs. Fry by immersion in hot lard, first testing it by throwing in a bit of bread, to see if hot enough. Serve immediately with either tomato sauce (page 155) or a plain white sauce (see next receipt). A circle of rice (boiled in milk) or boiled macaroni, or some flowerets of cauliflower, with the white sauce poured over both, is very good. Sweetbreads are often served with pease. The flavor of sweetbreads is much better if they are cooked to completion when once begun. It is not so well to parboil and allow them to get cold before frying.

Sweetbreads, with Cream Dressing, on Toast.

Boil a pair of sweetbreads as indicated in the last receipt, and, when they have been skinned and the pipes have been removed, cut them into good-sized dice. Then mix them in a sauce made as follows: Place in a little saucepan butter the size of a black walnut, and when it bubbles throw in a dessertspoonful (half an ounce) of flour; let it cook without coloring, then pour in gradually, stirring with an egg whisk, one and a half cupfuls of milk, or half milk and half cream; season it with salt and a suspicion of red pepper. This is seasoning enough for any invalid, yet sometimes a little nutmeg and sometimes grated cheese is also added. When the sauce is smooth, mix in the sweetbread dice, and when all is thoroughly hot, serve it immediately, poured over a well-made piece of buttered toast, partially moistened with a little hot water. Decorate the dish with parsley, or small leaves, or flowers of any kind.

Macaroni and Tomato Sauce.

Sauce: Put butter size of an egg into a saucepan, and when it is at the boiling-point throw in an onion (minced), two sprigs of parsley (chopped fine), and a little pepper. Let it cook five or eight minutes; then throw in a heaping tablespoonful of flour and a little broth from the stock-pot; if there be no broth, use a little boiling water; stir this well and let it cook five or eight minutes longer. Now pour in about a coffee-cupful of tomatoes which have been stewed and strained through a colander or a sieve, and stir all together. Boil half a pound of macaroni tender in well-salted boiling water or in stock, and drain it in the colander. Place alternate layers of the macaroni and the sauce on a hot dish, pouring the sauce over the top. Put the dish into the oven two or three minutes to heat. Serve immediately.

MACARONI AU GRATIN.

Ingredients: One cupful of well-boiled macaroni (macaroni added to well-salted water while boiling, and boiled about twenty minutes, or until soft, then drained); after it is chopped quite fine, one cupful of milk, two or three sprigs of parsley, or a heaping teaspoonful after it is chopped fine, a heaping teaspoonful of flour, one egg, butter the size of a black walnut. Put the butter in a little saucepan, and, when it bubbles, throw in the flour and cook it without coloring, then add the milk and the parsley; let it simmer a minute, then take it from the fire; add a little of the chopped macaroni to the egg for the purpose of beating it more easily, then add the sauce and remainder of the macaroni. Put it into a little pint pudding-dish or *gratin* pan, sprinkle over coarse bread crumbs which have been colored in a little butter, or place it in the oven for a few minutes to color the top, which makes it "*au gratin.*"

MACARONI CROQUETTES (Louis Bertholon, *Chef*).

Throw a third of a package (one third of a pound) of macaroni into salted boiling water, and boil it for twenty minutes; then cut it into quarter-inch lengths, forming little rings.

Prepare a sauce as follows: Make a *roux* by placing in a saucepan butter the size of a pigeon's egg; when bubbling, add a generous tablespoonful (a quarter of a cupful) of flour; let it cook a minute, and then add a cupful of stock, half a cupful of cream, two tablespoonfuls of grated cheese, fifth of a nutmeg (grated), salt, little pepper, and, when all is well mixed and cooked for a couple of minutes, take the mixture from the fire and stir in also the beaten yolk of an egg. Return the saucepan to the fire to cook the egg slightly, but do not let it boil, as that would curdle the egg. Now mix in evenly the macaroni rings (two cupfuls), and spread the mixture about half an inch in thickness over a pan. When cold it should be made into croquette form, egged and bread-crumbed, to be fried in boiling lard.

This mixture is quite soft to handle, but with a little practice it is easily managed. Take enough for a croquette with a spoon; shape it on the table with a knife; sprinkle over some sifted cracker crumbs, then lift it dexterously with a pancake turner on the plate of slightly beaten egg; turn it over with the pancake turner; then again lift it to a plate of sifted cracker crumbs. It can now be rolled without trouble.

Croquettes of all kinds are better to be quite soft.

Cheese served in this manner is not indigestible, according to M. Mattieu Williams in an article on "The Chemistry of Cookery," published in the *Popular Science Monthly*. Mr. Williams asserts, with good reasoning, that cheese, although indigestible when eaten raw, is very digestible when cooked and mixed with other articles of food. The diet is so hearty and rich, that when eaten in much quantity, other food should not be taken at the same time. In this receipt the cheese can be omitted if preferred.

These croquettes are to be served with tomato sauce.

TOMATO SAUCE.

Ingredients: One pint can of tomatoes; one sprig of parsley; half of a bay leaf; two cloves; one teaspoonful of onion, or one slice; salt and pepper. Add the seasoning to the tomatoes, and let them simmer all together for fifteen minutes, stirring occasionally. Pass it through the sieve, leaving out the seasoning. Place in a saucepan butter the size of a hickory nut, and, when it bubbles, add a teaspoonful of flour. Mix and cook it well, then add the tomato pulp, stirring it until it is smooth and consistent.

The sauce can be made one or two days before it is needed, if more convenient, and reheated just before serving.

A SALAD.

A lettuce salad is very wholesome served with meat. The usual dressing is a mixture of the following proportions: Three tablespoonfuls of oil; one tablespoonful of vinegar (a little less if the vinegar is quite strong); a saltspoonful of salt; half a saltspoonful of pepper; an even teaspoonful of onion, minced very fine. The salt, pepper, and onion are first mixed together, then the oil (by degrees), and lastly the vinegar.

A very good dressing for lettuce is furnished by any meat or fowl gravy (the thickened gravy better), and a very little vinegar.

OYSTERS ON TOAST, OR IN SHELLS OR PAPER CASES.

These oysters may be served on thin slices of toast, or in paper cases, or in shells, if convenient. A sprinkling of bread crumbs colored in a little butter would finish them in the paper or shells.

Put one quart of oysters (about twenty-five) on the

fire in their own liquor. The moment they begin to boil turn them into a hot dish through a colander, leaving the oysters in the colander. Put into the saucepan two ounces of butter (size of an egg), and when it bubbles sprinkle in one ounce (a tablespoonful) of sifted flour; let it cook a minute without taking color, stirring it well with a wire egg whisk, then add, mixing well, a cupful of the oyster liquor. Take it from the fire and mix in the yolks of two eggs, a little salt, a very little cayenne pepper, one teaspoonful of lemon juice, and one or two gratings of nutmeg. Beat it well; then return it to the fire to set the eggs, without allowing it to boil. Put in the oysters.

OYSTER CROQUETTES.

Place a pint of oysters (the measure nearly solid with oysters) over the fire, with a quantity of their cold liquor; when they begin to simmer, drain them quite dry from their liquor (through the colander), and cut them into large dice. If the oysters are small, cutting them into three or four pieces each will be sufficient.

Next, place butter size of a black walnut in a little saucepan, and, when it bubbles, throw in a dessertspoonful of onion, minced fine; let it fry a couple of minutes without taking color; then add a tablespoonful (quarter of a cupful) of flour; let it also cook a few moments without taking color; then pour in half a cupful of cream or milk, and half a cupful of the oyster liquor; season with salt, cayenne pepper (very carefully), and a few dashes of nutmeg. When it is evenly mixed and the flour is thoroughly cooked (a couple of minutes), take it from the fire, stir in the oysters, and set it away to get cold. Mould them, roll in egg (slightly seasoned with pepper and salt) and sifted cracker crumbs, and fry them by immersion into boiling lard.

They can be served with or without any of the sauces which are suitable for fish or meat; for instance, drawn butter sauce with either chopped hard-boiled eggs or capers mixed in; Bechamel sauce, the simple brown sauce, etc.

Serve them quite hot, directly from the fire.

SOME CREAM SOUPS.

It is perhaps a little troublesome to make the cream soups, as the material has to be passed through the sieve. They are exceedingly delicate and nourishing, however, and help to furnish a pleasant variety in a limited *repertoire* of dishes. The farina cream is especially simple. The cream of oysters is particularly good. I first saw it at Delmonico's, and wondered what could be the ingredients, admiring more than ever the consummate skill of those cooks.

The special enigma was, how the soup could be so light, as if raised with baking-powder. In learning how to make these soups afterwards, from a most able *chef* (Louis Cuppinger), it was a matter of surprise and satisfaction to find the oyster cream so simply made, containing only the ingredients of a common oyster soup.

The potato cream (*Purée Alexandra*) is delicious, and can be made without stock. Stock in itself contains some nutrition, and enough might well be made at once in winter to supply our invalid for a week.

The asparagus cream soup is also especially good.

For the oyster and chicken cream soups a small pestle and mortar (inexpensive) were considered by the *chef* indispensable for pounding the meat before passing it through the sieve. It is possible that after the meat is chopped very fine some other means may be suggested for pounding the meat, without coloring, if the pestle and mortar are not at hand.

A bowl of cream soup, with a couple of wafer crackers or a slice of Graham bread, might at times well constitute a sufficient meal for an invalid.

CREAM OF OYSTERS.

Put a quart of oysters with their liquor in a porcelain kettle or cleanest of saucepans over the fire. When the oysters are just about to boil, pour them into a colander (over a bowl), leaving the oysters in the colander. Chop the oysters as fine as possible, and pound them well in a mortar or thick bowl. Now make a *roux;* *i. e.*, put in a saucepan a piece of butter size of a small egg, and, when it bubbles, throw in a generous tablespoonful of flour (one and a half ounces); stir it well with the egg whisk, to cook the flour without allowing it to color; now pour in the oyster liquor, and when well mixed over the fire add the pounded oyster pulp and a pint of good cream. Pass this all through the sieve; season it carefully with salt and cayenne pepper; return it to the fire to heat without allowing it to boil, and, just as it is about to be served, add half a cupful of fresh cream, and a piece of butter size of a small pigeon's egg. Whisk it well with the egg-beater (keeping it hot, without boiling, over the fire) for a minute; pour into a warm tureen and serve immediately.

The *chef* sprinkled over the top some coarse, dry bread crumbs fried in a little butter. This addition is generally made to all the cream soups. Sometimes little fancy cuts of toast, cut with tin cutters, of diamond shape, are sprinkled over the top of the soup in the tureen, instead of the fried bread crumbs. For robust people little drops of fried fritter batter, looking like cooked beans, are sometimes sprinkled over the top of vegetable cream soups.

CREAM OF RICE OR FARINA OR BARLEY.

Put either a half cupful of rice or three fourths of a cupful of farina or barley into a quart of boiling clear stock, and let it cook until the grain is quite soft (about half an hour); then press it through the sieve, add two or three tablespoonfuls or more of good cream, and season carefully with red pepper and salt. Heat it again, and, just before serving, whip the soup in the tureen with the egg whisk.

CREAM OF CHICKEN.

When chicken is boiled for the family dinner (a sprig of parsley and a slice of onion being put into the kettle), a breast and some soft pieces of the chicken can be appropriated for our invalid. It should be chopped fine as possible, then pounded in a mortar, if one has it; and, if not, in a chopping-bowl. It is then moistened with a little of the chicken broth, and then pressed through a wire sieve. To a generous half cupful of this fine chicken pulp add about one cupful and a half of the chicken broth, free from fat. Thicken with a *roux;* i. e., in a little saucepan place a piece of butter size of a hickory nut, and, when it bubbles, throw in a teaspoonful of flour; let it cook without coloring; then add the chicken pulp and broth (mixed); stir well, and, when about to simmer, add a couple of tablespoonfuls of good cream, and a teaspoonful of parsley, chopped very, very fine. Season also with red pepper and salt. Whisk it with the egg-beater, before serving, keeping it hot, though not allowing it to boil.

CREAM OF ASPARAGUS.

This is one of the best of the cream soups. The receipt is given for two and one third quarts of soup, yet, of course, half or a third of the quantity can be made for the invalid, if more is not needed.

Ingredients: Two quarts of stock; about thirty stalks of asparagus; one half cupful of good cream; two tablespoonfuls of flour; butter size of pigeon's egg.

Boil the asparagus in the stock; cut and save some of the points, to serve in the soup; the remainder press through the sieve. Now make a *roux* by putting the butter in a saucepan, and, when it bubbles, throw in the flour, which cook a minute without coloring, stirring it well with the egg whisk. Now pour in the stock and the asparagus pulp, gradually at first; let it boil a minute, then add the cream, which heat, but do not let boil, for fear of curdling. Season to taste with salt and pepper. When the soup is in the tureen, ready to serve, sprinkle the asparagus points on top.

A SIMPLE ASPARAGUS SOUP (Dr. Comstock's Soup).

Fifteen or more stalks of asparagus are boiled in a quart of milk, and the whole (excepting some of the points) is passed through the sieve. It is then thickened with a *roux*, as in the preceding receipt, with butter size of a walnut, and a heaping teaspoonful of flour. A few tablespoonfuls of good cream can then be added, or it is very good without it, if it is not at hand. It is then seasoned to taste with salt and pepper, and served with the asparagus points sprinkled over the top.

CREAM OF POTATOES (*Purée Alexandra*).

Boil in water five medium-sized potatoes until they are nearly done; then pour off the water, and add a scant two quarts of clear stock, made with either veal or beef. When the potatoes are thoroughly cooked, pass them, with the stock, through a wire sieve; then add the beaten yolks of two eggs and half a cupful of good, thick cream; season carefully with salt and cayenne pepper. Stir it for a minute over the fire, to slightly cook the eggs, without allowing it to boil; then keep it at the side of the range (better kept in a double kettle or *bain marie*) until about ready to serve.

At the same time that the soup is being made prepare some vegetables for a garnish, as follows: Cut a medium-sized turnip (two ounces) into little dice this way—cut the turnip into slices about a quarter of an inch thick, without allowing the knife to cut quite through, so the slices will hold together; then slice them transversely in the same manner. Now, holding the turnip firmly together, cut off the ends into little dice about a quarter of an inch square. In the same manner cut a carrot (two ounces) into little dice; provide, also, a tablespoonful or more of pease and some string-beans cut into quarter-inch lengths.

All these vegetables may be used, or part of them, as convenient; the carrots and pease, however, are desirable for their fine color and flavor. Boil the vegetables separately, in little cups of salted boiling water; drain, and place them in the soup tureen. When about ready to serve, place the soup over the fire without allowing it to boil, and whip it vigorously with the egg whisk for one or two minutes; then add the vegetables, and serve immediately.

Or, the soup can be made without stock, boiling the potatoes in water, and adding more cream and a piece of butter size of a small egg.

CREAM OF STRING-BEANS.

Throw a quart of green string-beans in boiling water, in which there is half a tablespoonful of soda or as much carbonate of ammonia as would lay on the point of a knife, to preserve the color; drain the beans, and pass them through a sieve (not colander, but sieve). There will be about a pint of pulp. Make a *roux* by placing in a saucepan butter the size of a pigeon's egg, and, when it bubbles, throw in two large, heaping tablespoonfuls of flour (two generous ounces); let it cook without taking color; then pour in a quart of clear stock (see page 164), and the pint of string-bean pulp. Stir it well with the egg whisk, letting it cook a few minutes without boiling. It would be liable to curdle if boiled. Just before serving pour in nearly a cupful of good, thick cream; season with salt and cayenne pepper. Whip it well with the egg whisk over the fire, and serve immediately.

At Delmonico's they served, sprinkled over the soup in the tureen, imitation navy-beans, made by dropping drops of fritter batter in hot lard. They were crisp and savory, but a fritter of any kind should never be mentioned in an invalid's book.

CREAM OF CORN.

To a pint of grated corn (the sweet part, nearest the cob, well scraped) add a quart of hot water. Boil it for an hour, and press it through the sieve. Put into the saucepan butter the size of a small egg, and, when it bubbles, sprinkle in a heaping tablespoonful of sifted flour, which cook a minute without coloring, stirring well. Now add half of the corn pulp, and, when smoothly mixed, stir in the remainder of the corn; add a little cayenne pepper, salt, a scant pint of boiling milk, and a cupful of cream. Before serving, stir well with an egg whisk, to give it a light consistency.

Or, for a change, an addition to the soup of the yolks of two eggs, and the soup stirred a minute over the fire, although not allowed to boil, is good.

Or, a spoonful of chopped parsley may be added.

Cream of Corn (No. 2).

This is the *chef's* receipt. Place over the fire a pint of grated corn, with a piece of butter size of a walnut; let it cook only a minute, when pour in a quart of veal stock, and boil it an hour; pass it then through the sieve; add about three tablespoonfuls of cream; beat it again, and as it is about to be served stir it well with an egg whisk.

OTHER SOUPS.

Stock for Soup.

A good stock may be made by simply putting fresh lean beef or veal, with some bone, into clear, cold water (a gallon of water to three pounds of meat and bone), and let it simmer for five hours, passing it through the sieve, and seasoning it carefully with pepper and salt. It is better to make the stock the day before it is wanted, as then every particle of fat will rise to the top, and form in a hard cake, which can be removed at once, and the settlings can be avoided at the bottom, leaving a clear soup. There should never be a particle of fat left in a soup.

The flavor of the soup is much improved by a chicken addition. Occasion might be taken, at the time of making beef or veal stock, to have a boiled chicken for dinner, boiling it in the stock-pot. The flavor is also much improved by the addition of vegetables thrown in an hour before the stock is done. Four or five slices of onion, first fried (*sauted*) or colored in a little dripping on a platter before adding to the soup; also, the same quantity of sliced carrot, two good sprigs of parsley, and, if you have it, a stick of celery or a teaspoonful of celery-seeds, and a couple of cloves stuck in the onion. All contribute to the quality of the soup.

In winter enough stock ought to be made to last a week, as it will keep that time and longer in a cold place. Each day a portion of the stock jelly can be reheated, and, with different accompaniments, the invalid can have many changes. For instance, the addition of a few spoonfuls of cooked macaroni will make a good macaroni soup. A spoonful of cooked pease and other vegetables, in fancy shapes, would make a spring soup (or *Julienne*); a few spoonfuls of cooked tomatoes a tomato soup; toasted bread sippets, in fanciful shapes, a *potage aux croutons*. The stock, added to the cream soups, furnishes a dish for the most fastidious epicure, and a nutritious repast for the invalid.

In selecting the meat for soups cheap cuts from the leg and shoulder of beef are generally used. Ox-tails make good soup. Knuckles of veal, calf-heads, and tough chickens play a satisfactory *rôle* in stock. The meat, afterwards, can be made into palatable side dishes in the way of croquettes, etc.

I will give Gouffé's receipt for stock, showing the distribution of vegetables, as follows:

GOUFFÉ'S RECEIPT FOR STOCK OR BOUILLON.

Three pounds of beef; one pound of bone (about the quantity in that weight of meat); five and a half quarts of clear, cold water; two ounces of salt; two carrots, say ten ounces; two large onions, say ten ounces, with two cloves stuck in them; six leeks, say fourteen ounces; one head of celery, say one ounce; two turnips, say ten ounces; one parsnip, say two ounces.

Oyster Soup.

To one quart, or twenty-five oysters, add a half-pint of water. Put the oysters on the fire in their liquor. The moment it begins to simmer (not boil, for that would shrivel the oysters), pour it through a colander into a dish, leaving the oysters in the colander. Now put into the saucepan two ounces of butter (size of an egg) when it bubbles, sprinkle in a heaping tablespoonful (one ounce) of sifted flour; let the *roux* cook a few moments, without coloring; stirring it well with the egg whisk, add to it gradually the oyster juice, and half a pint or a cupful of good cream (which has been brought to a boil in another vessel); season carefully with Cayenne pepper and salt. Skim well, then add the oysters. Let it get hot without boiling and serve immediately.

Clam Broth.

This broth is much used of late years for invalids. Indeed, in New York it seems to be as standard a sickroom dish as beef tea. It can often be retained on the stomach when other foods disagree with the patient, and is a valuable substitute for milk, when that proves unsatisfactory. It is stimulating and nutritious. It can be administered by the spoonful, like beef tea, in cases of severe illness, or can be taken by the cupful, when, with a Graham cracker, it affords a hearty repast for others.

For half a pint (a cupful) use six large hard-shelled clams. Wash them well with a brush, and place them in a kettle with two or three tablespoonfuls of water over the fire.

The clam broth is simply the juice of the clam boiled for a minute. It does not require seasoning, as clam juice is salt enough; indeed, it has sometimes to be a little diluted with hot water to reduce the salt flavor. In pouring the juice from the kettle, avoid any particles of sand which may have settled at the bottom.

As soon as the clams are opened in the kettle they are sufficiently cooked; further cooking renders them tough. If

CLAM SOUP

is to be made, remove the clams from the shells as soon as they have opened, cut off the tough parts, and place them one side in a warm place, until the juice is prepared. Add about a cupful of hot milk to the juice, and thicken it with a *roux*, or a little flour. Now add the soft parts of the clams, bring the soup again to the boiling-point and serve.

Placing the live clams over the fire is a very cruel way to open them. Men-cooks and fishermen open them with a knife, a half-dozen in the course of half a minute.

FLOUR SOUP.

Put butter, size of a large hickory nut, into a little saucepan, and, when it bubbles, throw in a heaping tablespoonful of flour (a generous ounce). Stir it well with the egg whisk, allowing it to color evenly to a light brown. Take care that it does not burn. Now gradually pour in a pint of warm milk, stirring it well with the egg whisk. There should be no lumps. Let it cook for a minute only, when take it from the fire and add the beaten yolk of an egg. Return it to the fire for a few moments to set the egg, stirring well, and not allowing it to boil, as the egg would then curdle. Season with salt, a suspicion of red pepper, and a half teaspoonful of parsley chopped very fine. French cooks often add the same quantity of chopped cives, but the latter we will not recommend for our invalid.

It can be served with or without little toasts of bread, cut in thin slices and fanciful shapes before toasting.

The French and Germans often flavor "Soupe à la Farine" with a little sugar and cinnamon instead of salt, pepper, and parsley or cives.

DISHES OF RICE.

To Boil Rice.

For a teacupful of the boiled rice place a quart of clear water over the fire, and, when it boils hard, throw in two ounces, or two tablespoonfuls, of rice which has been previously well washed in cold water. Throw in also a teaspoonful of salt. Take off any scum that rises. In twenty minutes press some of the grains between the fingers, and if quite soft it is cooked enough. Do not cook the grains until they become broken. When done, pour the rice into a sieve to drain off the water; return the rice grains to the dry saucepan; cover them partly, and set them at the side of the fire to steam and dry.

To Boil Rice in Milk.

Bring one pint of milk to a boil, when stir in two tablespoonfuls of well-washed rice and a quarter of a teaspoonful of salt; pour it into a basin, cover it well, and place it in the oven to bake for an hour; or it may be cooked in the double saucepan. In a copper saucepan it could be boiled at the top of the range without burning, when it would be cooked in about twenty minutes.

Rice

May be served with many dishes; for instance, in a circle around chicken, fried (spring chicken) or boiled, or cold chicken dice stewed, with white or brown sauce poured over both the rice and chicken; or it may be served in the same manner with sweetbreads, or with stewed fruits, apples, peaches, pears, etc.

RICE AND GRAVY.

Fresh boiled rice wet with the juice from roast beef or mutton (free from fat) and served on a piece of toast.

RICE CONES.

Cook the rice in either milk or water, and while hot pour it into cups (which have previously been dipped in cold water) filling them about three fourths full. When cold and ready to serve, turn them out, arranging them uniformly on a platter; or, for our invalid, turn one into a small oval platter, or a saucer. Scoop out a little of the rice from the top of each cone, and put in its place any kind of jelly. Pour in the bottom of the dish a hot brandy sauce (see page 189), or hot sweet sauce of any kind, provided it is not flavored with vanilla.

A PLAIN RICE PUDDING.

The manner of making this most delicious and plainest of puddings was taught me by a most able *chef* (Louis Bertholon). The flavor is quite remarkable, considering that it is almost as simple as plain boiled rice.

For an invalid choose a little pudding dish holding about a pint. Put in a heaping tablespoonful of uncooked rice, fill the dish with boiling milk, and place it in the oven. Let it cook, stirring it once or twice (to prevent lumping) for about half an hour; then take it out and mix in a tablespoonful of sugar and half a teaspoonful of essence of lemon, or the thin, yellow cuts (without any white) of the rind of half a lemon, or with *fleur d'Orange*, or a sprinkling of nutmeg, or, indeed, any flavoring preferred, excepting vanilla, which is too unwholesome for invalids; return the dish to the oven, cooking altogether two hours, or one and a half hours, if the oven is quite hot. As the milk boils down more hot milk should be added (keeping the dish always filled) by lifting the skin and pouring in the milk at the side, or by removing the skin and allowing a new one to form. The dish will require about one and a half pints of milk.

RICE PUDDING (No. 2).

Another successful pudding, where every grain of rice lies in a creamy bed.

Ingredients: One cupful of boiled rice (better if fresh cooked and hot); three cupfuls of milk; three fourths of a cupful of sugar; one tablespoonful of cornstarch; two eggs; flavoring; or half these ingredients for a pint pudding dish. Dissolve the cornstarch first with a little milk, and then stir in the remainder of the milk. Bring this to a boil, when take it from the fire, and, when slightly cooled, stir in the  rice and the yolks of the eggs beaten well with the sugar. Return this to the fire (there is less risk of burning in a custard kettle) and stir until it begins to thicken like boiled custard, watching it carefully not to let it boil or curdle. Now, again, remove it from the fire, add the flavoring, say a scant teaspoonful of lemon extract, and pour it into a pudding dish. Spread over the top the whites of the eggs, beaten to a stiff froth, with a little sugar and flavoring added. Or, with the aid of a cone of writing-paper, decorate the top with a fanciful design, *à la méringue*. Give it a delicate color in the oven. To be eaten either hot or cold.

RICE À L'IMPERATRICE (Louis Cuppinger).

Place over the fire one pint and a half of milk, and  the thin yellow cuts of the rind of a lemon, and, when it boils, stir in half a teacupful of rice, and an even salt-spoonful of salt. When cooked (in about twenty minutes) stir in carefully half a cupful of sugar and a few drops of essence of lemon, or two or three spoonfuls of rum, or any flavoring. The rice should be rather moist when cooked. Spread it on a platter to get quite cold, then stir in carefully a half-pint of cream, whipped to a froth, and the fourth of a box of gelatine dissolved in a scant half cupful of water. To dissolve the gelatine, add it to the cold water, then set it for fifteen or twenty minutes in a warm place. Mould the rice. For the invalid it may be moulded in a teacup, or in one of the pretty little fancy moulds, which come of all sizes.

RICE PUDDING À LA GUILLOD.

Ingredients: a scant half cupful of rice; one pint of water; one cupful (half-pint) of milk; butter size of a hickory-nut; one tablespoonful of sugar; four eggs; salt; flavoring, say a scant teaspoonful of lemon extract, or two or three tablespoonfuls of rum.

When the water (salted) is at the boiling-point add the rice, and cook it twenty-five minutes; then add the milk (hot); cook it ten minutes longer; then add the butter, sugar, lemon, and well-beaten yolks of the eggs. Stir this for a few *moments* over the fire to set the eggs, without allowing it to boil. This batter can be stirred with a spoon for the purpose of partly breaking up the grains of rice, or it may be passed through a sieve; either way is very good. When the batter is entirely cold stir in dexterously the whites of the eggs beaten to a stiff froth, and put it immediately into a buttered double boiler (page 85), or into a long tin pail which can be covered, and set into a pot of boiling water, the water reaching about three fourths to the top. A weight should be placed on top of the tin pail to keep it from turning. Cook about three quarters of an hour. Turn out carefully on a platter, and serve with currant or plum jelly sauce.

This receipt is made by my most able cook, Louise Guillod, who for six years has relieved me of all responsibility of the *cuisine*.

Currant or Plum Jelly Sauce.

Stir two dessertspoonfuls of currant jelly (a scant third of a cupful) and two tablespoonfuls of sugar into one and a half cupfuls of cold water. It is sometimes difficult to dissolve the jelly. Bring it to a boil, then add a teaspoonful of either cornstarch or flour for a thickening, first rubbed smooth in a little cold water; let it cook two or three minutes. To be served cold.

A Rice Dish (to be served as a vegetable).

Mix carefully (not to break the grains) in a pint of boiled rice a tablespoonful of either minced parsley or cives; put a piece of butter size of a pigeon's egg into a saucepan, and let it color a light brown; mix the rice in the butter, and serve hot as a vegetable. A little mound of this rice may be placed in the centre of a platter, with a row of green pease around it.

CREAMS AND FRUITS.

Whipped Cream.

There is no more wholesome, nutritious, and delicious dessert for an invalid than whipped cream, either served simply with a wafer biscuit or some very thin slices of sponge cake (cake not very allowable) around to form a *charlotte-russe*, or with a stewed pear, peach, apple, or some wine jelly for a centre.

No better ice-cream can be made than the simple one of whipped cream frozen. The cream (thick) is sweetened and flavored with any of the flavoring extracts (except vanilla) or any of the sweet wines or liquors. It is delicious merely sweetened. The cream froths more readily when quite cold. The cream-whipper is recommended, yet, without this, vigorous whipping with a silver fork will accomplish the result. The froth, as it rises, is to be skimmed off and placed on a sieve; that in the dish below is returned to be rewhipped. Place the cream froth, as soon as all whipped, on the ice, to remain until served.

Clabbered Milk (Dr. Gatchell).

Set a quantity of skimmed milk away in a covered glass or china dish. When it *turns, i. e.*, becomes smooth, firm, and jelly-like, it is ready to serve. Do not let it stand until the whey separates from the curd, or it will become acid or tough. Set it on the ice for an hour before wanted for use. Serve from the dish in which it has turned. Cut out carefully with a large spoon, put in saucers, and eat with cream and nutmeg. This is one of the most wholesome of dishes, and those to whom it is new soon acquire a taste for and grow fond of it. To be relished the clabber must be new and fresh. It is generally eaten with cream, sugar, and sometimes nutmeg.

The clabbered milk is an excellent diet for some dyspeptics; however, they should not eat it with sugar. In case the clabbered milk is not agreeable at first, begin with a small quantity—a tablespoonful at a time—and gradually a taste for this very useful food can be acquired. Some prefer taking it as a drink, beating it up until it becomes creamy.

Cottage Cheese.

Cottage cheese is made of the curd left after separating the whey from clabbered milk.

Tie the clabbered milk in a cloth, hang it (for instance overnight) and let the whey drain out.

Or, place a pan of clabbered milk over a kettle of boiling water until the whey becomes merely hot. If the pan is placed directly on the range, let the whey become merely hot and no more. The boiling-point would spoil the cheese by making it tough. The whey is then pressed from the curd and the latter is mixed with cream or butter, or both, and salt, making the cheese rather moist, yet firm enough to mould into balls.

Ice-cream and Iced Peaches.

Sometimes ice-cream is given to a patient suffering with a fever or inflammation of the stomach. The simplest and richest ice-cream is pure cream sweetened to taste, and flavored with *fleur d'orange*, extract of lemon, or a very little sherry (never vanilla), and frozen. It is still better to whip it and freeze the whipped cream. If it is desirable that the cream be not so rich, a simple frozen boiled custard is very good.

The custard is made by adding the yolks of two or three eggs, well beaten, with a tablespoonful of sugar to a pint of fresh milk. This is stirred in a double boiler, or in a tin pail placed in a second vessel containing boiling water, until it just begins to thicken. It is then removed at once (to prevent curdling) and seasoned as just described for whipped cream. The iced custard is improved by stirring in it, when partly congealed in the freezer, more or less whipped cream. However, this adds again to its richness. Chopped peaches or grated pineapple could also be added at this time.

A mixture of sweetened fresh peaches, pared, stoned, and quartered, with or without cream mixed with them, and frozen in a mould (without stirring the mixture) is also a most delicious dish for a febrile sufferer. I have never tasted canned peaches or pears frozen in this way, but think they might be satisfactory if the fruit were especially good.

BAKED APPLES, ETC.

Nothing is more simple, wholesome, and palatable than a baked apple served with cream and sugar. The canned peaches are generally heavy for an invalid; and, by the way, tin-canned tomatoes and acid fruits are forbidden entirely by many physicians, the tin having a deleterious effect on the acid of the vegetable or fruit.

Baked apples are prepared as follows: With a sharp-pointed knife, or an apple-corer, remove the cores without breaking the apples. Set them in a pan just large enough to hold them. Fill the apertures with sugar, and for a change a small stick of cinnamon, or the thin, yellow slices of lemon rind can be inserted also. Pour a half cupful of water into the pan and bake the apples until tender. They are oftener cooked without cover, yet are very good covered with a basin and allowed to cook in the steam. For a change the apples may be pared. Serve with cream and sugar. Baked apples and stewed prunes are probably the most wholesome sweetmeats for an invalid, and can be served at any meal.

APPLE SAUCE.

Apples (pippins especially good), neatly and evenly quartered (having already been pared and cored) are placed in a porcelain pan with enough cold water to barely cover them. Sugar to taste is added, and perhaps some raisins, sometimes slices of lemon, sometimes a few sticks of cinnamon, each or all, are added. Cook them slowly, and the moment the apple quarters are tender when pierced with a fork, they are done, ready to be poured into some pretty glass dish, and allowed to get cold before serving.

Sometimes the apple is stirred into a half *purée*, or pulp, and sometimes it is passed through the sieve.

A good apple sauce is made by adding to the apple which has been passed through the sieve, and sweetened to taste, the beaten whites of eggs just before it is served—say the whites of two eggs stiffly beaten, to a pint of apple pulp.

PEAR OR OTHER FRUIT COMPOTE.

A compote is merely the fruit (pear, peach, apple, plum, etc.) boiled whole with only enough water to cover it, and sweetened to taste. The fruit is only cooked until tender. Pears are generally selected for compotes when not quite ripe.

The California dried pears, stewed until tender, and

sweetened to taste, are most excellent for our invalid when the fresh pears cannot be obtained.

Compotes are often served with a circle of rice (boiled in milk) around, or the rice may be in the centre and the fruit placed around it.

Sea-moss Blanc-mange.

Wash one and a half ounces of Iceland or Irish moss in cold water, then place it over the fire in a cupful (one half pint) of fresh, cold water. Stir it occasionally until soft; add then one and a half cupfuls of warm milk and three lumps of sugar. Place the little saucepan containing these ingredients into a second larger saucepan half filled with boiling water, and let the water boil until the moss is entirely dissolved. Pour this into teacups or little moulds previously wet with cold water. Turn them from the moulds when hardened and ready to serve, and serve each mould with three or four tablespoonfuls of cream poured around, and, perhaps, a preserved strawberry half buried on top; or a fruit compote of any kind can be poured around the blanc-mange.

Cornstarch Blanc-mange.

Allow three tablespoonfuls, or three quarters of a cupful, of Duryea's cornstarch to a quart of milk. Stir enough of the cold milk into the cornstarch to make a soft, smooth paste; bring the remainder of the milk to the boiling-point, stir in the paste, and boil it about three minutes, taking care that it does not burn. Pour it into cups or moulds previously wet with cold water, and set it in a cold place to harden. Serve with sweetened cream or a little soft-boiled custard, and a couple of preserved strawberries for a garnish.

CHARLOTTE-RUSSE.

The charlotte-russe made after this receipt is undoubtedly one of the very best ever made.

Bring a cupful, or half a pint, of milk almost to the boiling-point, and then stir in the yolks of four eggs, previously well beaten, with three tablespoonfuls of sugar. Stir this carefully over the fire (in a double kettle) making a boiled custard. Care must be taken that it does not curdle, or become too thick. Take it from the fire and add to it a quarter of a boxful of gelatine, previously soaked with enough milk to cover it in a cup, and dissolved, by setting it at the side of the fire. Add also, when the custard is a little cooled, two or three tablespoonfuls, of best sherry wine for a flavoring. Set this custard on ice, or in a cold place, until partly congealed, and then stir into it, evenly and carefully, a quart of cream whipped to a stiff froth.

This can be poured into either a charlotte pan, or little paper cases (page 153), lined with lady-fingers, or into a pretty glass dish with a row of lady-fingers around the sides, and then it is served in the same dish.

If sponge cake is objectionable for the invalid, the creamy custard, which is simple and wholesome enough for almost any one, can be served alone, in the paper cases.

CUSTARDS.

Plain Baked Custard (very good).

A PLAIN custard may be made with a pint of milk, either two whole eggs or the yolks of three eggs, and a couple of tablespoonfuls of sugar. It can be flavored with a little nutmeg or extract of lemon. It is very good without flavoring. The eggs and sugar are well beaten together before the milk is added. It is poured into a small pudding dish or basin, and this is set in a larger basin containing hot water, which reaches three fourths to the top of the pudding dish. The two vessels, one in the other, are then placed in the oven until the custard is set (about twenty minutes). As soon as it is *set* it is done, and should not be left to allow the whey to separate. This is the very best way to bake custards.

Custard à la Morrison (a delicious custard).

Make a boiled custard with a pint of milk, the yolks of three eggs (if small), and a tablespoonful of sugar.

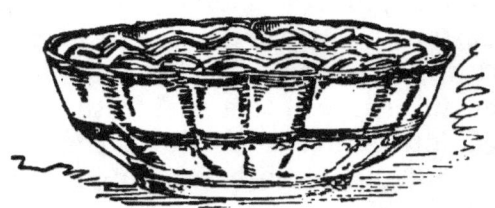

The yolks and sugar are beaten together, the milk added when warm, and the whole cooked in the double boiler. It must be stirred constantly while cooking, and the instant watched when it is of exactly the right thickness, resembling rather thick cream. If allowed to remain a moment too long it curdles and is spoiled. A *chef* tells me, however, that if a custard or *purée* soup begins to curdle it can be stopped by pouring in quickly a little cold milk or water, and stirring very regularly for a few minutes. When the smooth boiled custard is *cold*, and flavored with anything but vanilla, the whites of the eggs, beaten to a stiff froth, are mixed in smoothly with the egg whisk.

The top of the custard may be decorated with a little of the egg froth mixed with a little bright red jelly, with the aid of a paper funnel or *méringue* decorator, or the white, for decorating, may be stirred with zest, or thin slices of lemon peel (without white), and slightly sweetened. This will give a delicate green color to the *méringue* as well as a delicious flavor. The lemon strips are to be removed. The custard should be served soon after the beaten white of the egg is mixed in, as the egg froth is not cooked.

Tapioca or Sago Custard

is merely an addition to a plain custard (before it is baked) of more or less tapioca or sago after it has been soaked an hour or more in hot water.

The two following are from Gouffé's "Receipts for the Sick," called by him "*Petit pot de crême, au café,*" and "*Au chocolat.*" It may not taste as well under the common name of

A Cup of Coffee Custard.

Beat well in a coffee cup or small fancy pudding dish the yolks of two fresh eggs and a teaspoonful of sugar. Then mix into it four tablespoonfuls each of fresh-made, clear coffee, and milk. Set the cup into a basin of hot water so that the water will reach nearly to the top of the cup; put this into the oven and cook about fifteen minutes, or until the custard is set without curdling. To be served hot or cold.

A Cup of Chocolate Custard.

Put a heaping teaspoonful of grated chocolate with two tablespoonfuls of milk, and stir it over the fire until perfectly smooth; then add six tablespoonfuls of rich milk, and also the yolks of two eggs which have been well beaten, with a teaspoonful of sugar. Cook the same as coffee custard, and serve either hot or cold.

Granulated or Crushed Barley, Oat, or Wheat Custard.

The grain is thrown into salted boiling water and cooked fifteen or twenty minutes, or until thoroughly done. It is then drained, and a few tablespoonfuls (the custard should not be too thick with the grain) are added to a plain baked custard (page 180), before it is baked. Or the cooked grain can be substituted for rice in rice pudding No. 2 (page 171).

Rennet Custard.

A very palatable and digestible dish for an invalid.

Sweeten some milk to taste; place it over the fire until lukewarm; remove it from the fire and mix in it thoroughly some liquid rennet (it comes prepared for custards, and can be purchased at the druggist's), in the proportion of a tablespoonful of rennet to a quart of milk, in summer—perhaps a very little more rennet in winter. Let the milk stand lukewarm until a quite consistent curd is formed, then put it in a cold place until served.

The milk should be prepared in the dish in which it

is to be served; for, if it is disturbed, the whey will separate, which must be avoided. It is served with a little cream, or whipped cream poured over, and perhaps garnished with a preserved strawberry or two on top.

Sometimes the milk might be flavored with a very little brandy, rum, curaçoa, or maraschino before the rennet is added.

Caramel Custard.

Make the caramel by putting two tablespoonfuls of brown sugar and a teaspoonful of water over the fire and stirring it until it gets a quite dark brown—not black; then add a dessertspoonful of water. It will make a thick syrup. Pour this into the bottom of two cups or little fancy moulds, and turn it around until it covers the bottom and sides.

For the custard, beat well three eggs (yolks and whites), with a teaspoonful of white sugar and the very thin yellow cuts of a lemon; then stir in a cupful of milk or thin cream which has been brought to the scalding-point (not boiling) over the fire.

Fill the cups or moulds (previously lined with the caramel) with the custard; place them in a basin of hot water, the water reaching nearly to the top of the moulds, and bake them in the oven until the custard is set, or feels firm to the finger—no longer. They will set in twelve or fifteen minutes. The custards may be served either hot or cold—although they are generally served cold—turned from the mould when just ready to be served.

JELLIES.

Wine Jelly.

INGREDIENTS: One box of gelatine, soaked in one pint of clear, cold water, one pint of wine, the juice and the thin cuts of the rinds of three lemons, one and a quarter pounds of sugar (or according to taste), one quart of clear boiling water, the whites of two eggs (well beaten) and the shells, and a small stick of cinnamon.

Soak the gelatine in the pint of cold water an hour, then pour over it the quart of boiling water, stirring it well; now add the wine, sugar, lemon juice (strained in a fine strainer), and the thinnest possible cuts from the peels of the lemons. These cuts take only the little globules of oil in the peel, which are exceedingly delicate in flavor, the white being bitter. Add, also, the small stick of cinnamon, as it adds much to the flavor of the jelly. Put this into a porcelain kettle; let it boil rapidly about a quarter of a minute without stirring it; now, setting the kettle on the hearth, let it remain another half-minute to settle; then skim off carefully the scum which is on the top, and pour it through the jelly-bag. It should be entirely clear; if, however, the first should not be so, return it to the bag. Cold water should be poured into the moulds, then emptied just before using. Jelly hardens much quicker on ice, or in the coolest place to be found. Dip the moulds into warm water a moment before taking out the jelly. If allowed to remain a moment too long the jelly might dissolve too much and injure the form.

Many kinds of wines and liquors may be used. The above receipt is well-proportioned for sherry, champagne, madeira, or port. A smaller proportion of brandy, maraschino, noyau, or of punch would make sufficient flavoring.

Wine Jelly (without cooking).

Ingredients: one pint (two cupfuls) of wine; one pint of sugar; one pint of cold water; one pint of boiling water; half a package of gelatine; two lemons.

Add the gelatine to the clear, cold water, and let it soak for an hour or more; dissolve the sugar in the hot water, and when it is boiling stir it into the soaked gelatine; add the strained lemon juice and the thin, yellow cuts of the peel, and, when all is dissolved, the wine. Strain through a flannel cloth or bag or a napkin, without pressing it. If in a hurry for the gelatine, it will dissolve quicker if set in a warm place. In hot weather it is advisable to use a little more gelatine than in cold weather, or as in receipt. If brandy or rum is used, half the quantity mentioned for wine would be taken, or enough could be poured in to suit the taste.

Coffee Jelly.

Soak three quarters of a box of gelatine (either Cox's or Cooper's, or ten sheets of the common gelatine) in a pint of cold water until dissolved; then add a pint of boiling water, two cupfuls of sugar, and one pint of clear, *strong* (so the *chef* said) coffee. But the coffee need not be so very strong. Mould it. Surround coffee jelly, when on the platter ready to be served, with whipped cream.

Currant Jelly.

Pick out the leaves from the currants, but it is not

necessary to be particular about all the stems. Mash the currants with a potato masher, and cook them enough to merely free the juice, without adding any water. Strain the juice, and allow one pound of sugar for one pound of juice. Boil the juice fifteen minutes after measuring it, and then take it from the fire, and add the sugar, allowing it to dissolve without further boiling or cooking of the juice. When the sugar is well dissolved and mixed in the juice, pour it into glasses. Fasten over the covers when the jelly has hardened.

Currants should not be picked just after a rain.

Currant Preserves.

Allow one pound of sugar to one pound of currants. Free the currants from the stems, and cook them fifteen minutes; then add the sugar and a few raisins, and, as soon as it comes to the boiling-point again, seal them tightly in glass jars.

Orange Marmalade.

This marmalade furnishes one of the best and cheapest comfitures which can be made in the large cities, and a very little of it, used for garnishing a blanc-mange, etc., or for spreading on bread-and-butter, is not unwholesome for a convalescent. It is made in January or February, when oranges are cheap, and the expense will not be over fifteen or twenty cents a glass.

Allow one lemon to six oranges. Quarter the skins, and boil them slowly two hours and a half; then scrape out the soft pulp from the inside to be thrown away, and cut the outside skins into shreds. Squeeze all the juice possible from the fruit; weigh the juice and skin shreds together, and allow three fourths of a pound of sugar to a pound of fruit. When the fruit and sugar are mixed, let them simmer for an hour. If one prefer, the whole pulp of the fruit may also be added. It does not make so. clear a preserve, yet it is added in the Dundee marmalade.

STRAWBERRY PRESERVES.

Allow three fourths of a pound of sugar to a pound of fruit. Let the sugar simmer twenty minutes, adding perhaps a tablespoonful of water to start it; then add the strawberries; let them come merely to a boil; then cover, and place them at the back of the range to steam five minutes. Put them into glass jars while still scalding hot, and seal them hermetically.

PUDDINGS, ETC.

Corn Cottage Pudding.

Ingredients: One cupful of cornmeal flour; half a cupful of sugar; one cupful of milk; one tablespoonful of lard (size of small egg); three eggs; one teaspoonful of baking-powder; a little salt.

Mix the baking-powder and salt well into the flour, then add the sugar and yolks of the eggs well beaten together), the lard (melted), and lastly the milk, and the whites of the eggs which have been beaten to a stiff froth. Mix this smoothly, pour it immediately into a buttered round tin basin and bake about twenty minutes. Take care to have the cake baked just in time to be served. It is to be eaten hot with a liquid sauce. The following is a simple one.

Plain Pudding Sauce.

Ingredients: One pint of water (two cupfuls); three fourths of a cupful of sugar; a piece of butter the size of a walnut; a tablespoonful of either cornstarch or flour; flavoring of either brandy, rum, lemon, or wine (with or without a little nutmeg), or zest and cinnamon.

When the water boils, stir in the cornstarch or flour (rubbed smooth with a little cold water) and also the sugar. Boil it well for four or five minutes, to thoroughly cook the cornstarch or flour. Take it then from the fire, and stir in the butter and flavoring.

This is a good-enough plain sauce; it is improved,

however, by adding the well-beaten whites of one or two eggs, and stirring it well with the egg whisk for a minute over the fire to set the egg and make the sauce quite smooth.

GRAHAM-FLOUR PUDDING.

Ingredients: One and a half cupfuls of Graham flour; half a cupful of molasses; a fourth of a cupful of butter; half a cupful of sweet milk; one egg; an even teaspoonful of soda; three quarters of a cupful of English currants, or raisins (or mixed).

Into the flour pour the molasses, the butter partly melted, the egg (beaten), and the fruit. Mix all evenly together, then add the soda, dissolved in the milk. Steam two and a half or three hours.

A double tin pail (see cut, p. 85) is best adapted for steaming. The water in it should be boiling when the pudding is first placed in it, and when it needs replenishing *boiling* water should be added, so that it should at no time stop boiling. Serve with plain sauce (see page 188).

FARINA PUDDING.

Ingredients: One pint of milk; three quarters of a coffee-cupful of farina; half a cupful of sugar; butter the size of an egg; the thin yellow cuts of the peel of a lemon; four eggs.

When the milk is just boiling add the farina, and after it has cooked a few minutes stir in the sugar, lemon peel, and butter; let it cook slowly half an hour, then take it from the fire, and, when slightly cooled, stir in smoothly the yolks of two eggs. Take out the lemon strips. When the mixture is *quite cold*, stir in lightly the whites of the four eggs, beaten to a stiff froth, and put it in a high mould or long tin pail, prepared as follows: Butter the inside with a glazing brush, throw in a handful of sugar, and leave in the mould all the sugar that will stick to the sides; then add the pudding and place the mould in a basin of water, the water reaching about half or three quarters to the top of the mould. Let it cook (*au bain marie*) on the top of the range for ten minutes; then put all (basin of water as well) in the oven to bake for an hour. Serve immediately with currant-jelly sauce or Sauce Burke (page 191).

Quiogue Pudding.

Ingredients: Five of the ordinary Boston soda crackers, or three fourths of a cupful when rolled; a quarter of a cupful of flour; two eggs; a generous half cupful of milk.

Roll the crackers, stir in the milk, then the flour, and eggs (beaten separately). Cover it tightly in a mould or small tin pail, and boil it half an hour in a large vessel of boiling water. Serve with a hard sauce of butter and sugar rubbed to a cream with nutmeg sprinkled over, or with Sauce Burke or a currant-jelly sauce. However, any of the pudding sauces will answer.

Macaroni Pudding

is merely a baked custard pudding (page 180) with a quarter or half as much fresh, boiled macaroni added as there is custard. Of course the macaroni is added before the custard is baked.

Fine Granulated-wheat Pudding.

Ingredients: a scant half cupful of the wheat; one cupful of milk; two eggs; butter size of a small hickory nut; pinch of salt.

Bring the milk to a boil, then add the wheat and salt, and cook about five minutes. Take it from the fire, and add the yolks (beaten) and the butter. Let it get quite cold, then add the whites of the eggs, beaten to a stiff froth. Place it immediately in the oven, to cook about twenty minutes.

In cooking all *soufflé* puddings the oven should be hot, and for the first two or three minutes after the pudding is in, the oven-door should be slightly opened, so that the pudding can become evenly heated through, before it begins to rise. The pudding can be served with or without a sauce; however, a sauce is an improvement, and the following might be selected.

SAUCE BURKE (a delicious pudding sauce).

Bring a pint of milk to the boiling-point, and then stir in a generous teaspoonful of cornstarch, previously rubbed smooth with a little of the cold milk; add also a tablespoonful of sugar. Let it cook for two or three minutes to thoroughly cook the starch, and then let the mixture get entirely cold. Flavor it with sherry or any of the flavorings, and just before serving stir in evenly the whites of two eggs beaten to a stiff froth. As the egg froth is not cooked, the sauce will not keep very long at its best, perhaps half an hour.

SAUCE GUILLOD.

Whip the whites of two eggs to a very stiff froth; the froth of one egg should more than fill a goblet if properly whipped. In a small saucepan put two tablespoonfuls of granulated sugar, with two tablespoonfuls of water; let it cook without stirring for three or four minutes, or until it forms a syrup, not quite thick enough to candy. It must be watched carefully, then add the egg froth, which stir in with an egg whip quite vigorously for a minute at the side of the fire. Stirring will give the froth a fine grain. Take it from the range and add enough fresh lemon juice to take away the excessive sweetness of the *méringue*.

OTHER SOUFFLÉ OR PUFFED PUDDINGS.

The last-named pudding (fine granulated wheat) can be made as well with rice, farina, granulated oats, granulated barley, etc. It is especially good made with crushed barley. The barley must be well boiled in water (twenty minutes) before it is added to the milk, etc.

BARLEY PUDDING (simple).

Ingredients: Two cupfuls hot milk; half cupful of barley; one tablespoonful sugar; a pinch of salt. Into the pint of hot milk stir the barley. Season with a pinch of salt; add a tablespoonful of sugar, and place it in the oven for about twenty minutes; stir it occasionally until the barley is swelled, then add half a cupful of extra hot milk and let it bake slowly for an hour.

ORANGE PUDDINGS À LA MUTREUX.

Soak a cupful of stale bread in half a cupful of milk until it can be beaten to a pulp; mix with it the grated rind of one orange, the juice of two, sugar to taste, and the yolks of two raw eggs; butter six small cups, and set them in a pan of hot water; then beat the whites of two eggs to a stiff froth, mix them lightly with the other ingredients, partly fill the cups, and bake the puddings until the egg is done, in a moderate oven: about fifteen or twenty minutes will be required; serve the puddings hot.

LEMON (health-food) PIE OR PUDDING.

For two pies, rub until smooth two heaping tablespoonfuls of granulated wheat or barley and one tablespoonful of cornstarch (a scant three quarters of a cupful all together) with six tablespoonfuls (a scant half cupful) of cold water. Add to this two cupfuls of boiling water, and let it simmer over the fire three or four minutes, until the flour is thoroughly cooked. Take it off the fire, and when partly cooled add the yolks of three eggs, beaten with one and a half cupfuls of sugar to a froth, a piece of butter the size of a black-walnut, and the grated rind and juice of a large lemon. Bake with under crusts, and when done spread over the top the beaten whites of three eggs, with a heaping teaspoonful of sugar added (after they are beaten), and color in the oven.

The pie is much more attractive if the *méringue* is put on in fancy design, with a paper funnel (made of thick writing-paper and a pin) or the *méringue* decorator. The egg froth should be slightly sweetened, and flavored by stirring in the yellow cuts of lemon peel, which are afterwards removed. The lemon peel gives delicate flavor as well as color to the *méringue*.

The pie-paste can be made more wholesome by using very little lard or butter and a small portion of baking-powder. Or, the paste may be made with half Graham flour (sifted) and half white flour, a little baking-powder, and mixed with cream. The crust may be rubbed over with a little of the beaten white of an egg before the custard is added, which will prevent it from soaking into the crust. The custard may be baked in a little pudding dish without pie crust.

Graham Sponge Cake.

Ingredients: six eggs; three cupfuls sugar; four cupfuls flour (sifted Graham flour recommended); one cupful of cold water; two teaspoonfuls of baking-powder; juice and grated rind of half a lemon; a little salt.

Mix the yeast powder and salt well into the flour,

sifting it once or twice; stir the yolks and sugar to a froth; add first to the flour, etc., the yolks and sugar, and then the egg whites (beaten to a stiff froth), and then the lemon and water. The materials should be all ready, viz., the pans buttered, the flour and sugar sifted, the lemon grated, strained, etc., so that no time will be lost in mixing them together and getting them quickly into the oven.

For robust persons a sponge cake is often covered with a wafer thickness of icing, made by stirring a heaping cupful of pulverized sugar into the white of an egg (not previously beaten), and flavored with lemon, vanilla, or rum, etc.

BILLS OF FARE FOR CONVALESCENTS.

The following bills of fare are given for the purpose of suggestion, although the diet is a hearty one and only calculated for patients taking a certain amount of exercise and requiring a generous diet.

An invalid confined to the bed should be satisfied with very little sweets, and a breakfast or tea consisting only of an oatmeal or farina porridge and cream, cracked wheat and cream, a slice of Boston brown bread, or toasted Graham bread and cream, cornmeal mush and milk, rice and milk, poached egg or raw egg, a plain dish of macaroni, a cream soup, any of the gruels, a custard with Graham bread, a *soufflé* pudding of barley, granulated wheat, Graham flour, etc., any of the rice dishes, or other single dish as simple and nutritious as these, with a simple accompaniment of bread and apple sauce, or fruit compote, and a cup of hot water, grape juice, or fresh koumiss for a beverage.

BREAKFAST (at 8 o'clock).
Cracked Wheat Mould and Cream,
Bread Sippets,
Cup Hot Water with Sugar and Cream (better than tea or coffee.)

DINNER (at 1 or 2 o'clock).
A Slice of Rare Roast Beef, or Broiled Beefsteak,
A Baked Potato, Apple Sauce,
A Chocolate Custard.

TEA (at 6 o'clock).
Rice Cone with Hot Sauce,
Graham Bread, Grape Juice.

BREAKFAST.
A Slice of Boston Brown Bread with Cream poured over,
A Poached Egg on Toast,
Cup of Hot Water.

DINNER.
A Fricassee of Chicken, Potatoes *à la Crême*,
Lettuce dressed with the Sauce of the Fricassee and a few drops of Vinegar,
Graham-flour Pudding, Sauce Burke.

TEA.
A Small Dish of Macaroni and Tomato Sauce,
A Pear Compote.

BREAKFAST.
Oatmeal Porridge,
Oysters on Toast,
Cup of Chocolate.

DINNER.
A Lamb or Mutton Chop with Mashed Potatoes,
Spinach on Toast,
Macaroni Pudding.

TEA.
Cornmeal Mush and Milk.

BREAKFAST.
A Chicken Croquette with Pease around,
Milk Toast of Graham Bread,
Cup of Hot Water.

DINNER.
Cream of Asparagus, or Rice, Barley, etc.,
Boiled Fish, Carrots à la Crême,
Baked Potatoe, a Banana, Grape Juice.

TEA.
Barley Pudding, Sauce Burke,
Cup of Hot Water.

BREAKFAST.
A Sweetbread with Rice around, Cream Sauce,
Oatmeal Porridge.

DINNER.
Boiled Chicken and Macaroni,
Stewed Corn,
Farina Pudding.

TEA.
Hard Graham Rolls,
Grape Juice,
Custard à la Morrison.

BREAKFAST.
Boiled Eggs,
Baked Apple and Cream.
Corn Bread.

DINNER.
A Breast of Prairie Chicken, Mashed Potatoes,
Stuffed Tomatoes,·
Corn Cottage Pudding.

TEA.
Cracked Wheat and Cream.

Breakfast.

Farina Porridge, Fried Mush and Sugar Syrup,
Fruit Compote.

Dinner.

Slice of Roast Mutton, Salad,
Potatoes à la Neige,
Rice Soufflé (Pudding à la Guillod).

Tea.

Cup of Chocolate,
Granulated Wheat Pudding,
Stewed Prunes.

Breakfast.

Raw Egg (whipped),
Cornmeal Pancakes, Sugar Syrup,
Sweet Oranges Sliced.

Dinner.

Clear Soup with Bread Dice,
Fried Spring Chicken with a surrounding of Rice or Cauliflower and Cream Sauce, String-beans,
Rice à la Imperatrice.

Tea.

Macaroni Croquettes, Tomato Sauce,
Graham Bread,
Grape Juice.

APPENDIX.

Extract from an Article on the Effects of Tea and Coffee on the System, also on Count Rumford's Substitute for Tea, by M. Mattieu Williams.

(Published in *Knowledge*; republished in *The Popular Science Monthly* of December, 1884.)

"TAKE eight parts by weight (say ounces) of meal (Rumford says 'wheat or rye meal,' and I add, or oatmeal), and one part of butter. Melt the butter in a clean *iron* frying-pan, and when thus melted sprinkle the meal into it; stir the whole briskly with a broad wooden spoon or spatula till the butter has disappeared and the meal is of a uniform brown color like roasted coffee, great care being taken to prevent burning on the bottom of the pan. About half an ounce of this roasted meal, boiled in a pint of water, and seasoned with salt, pepper, and vinegar, forms 'burned soup,' much used by the wood-cutters of Bavaria, who work in the mountains far away from any habitations. . . . The rye bread, which eaten alone or with cold water would be very hard fare, is rendered palatable and satisfactory, Count Rumford thinks also more wholesome and nutritions, by the help of a bowl of hot soup, so easily prepared from the roasted meal. He tells us that this is not only used by the wood-cutters, but that it is also the common breakfast of the Bavarian peasant, and adds that 'it is infinitely preferable, in all respects, to that most pernicious wash, *tea*, with which the lower classes of the inhabitants of Great Britain drench their stomachs and ruin their constitutions.' He adds that, 'when tea is taken with a sufficient quantity of sugar and good cream, and with a large quantity of bread-and-butter, or with toast and boiled eggs, and, above all, *when it is not drunk too hot*, it is certainly less unwholesome; but a simple infusion of this drug, drunk boiling hot, as the poor usually take it, is certainly a poison, which, though it is sometimes slow in its operation, never fails to produce fatal effects, even in the strongest constitutions, where the free use of it is continued for a considerable length of time."

"This may appear to many a very strong condemnation of their favorite beverage; nevertheless, I am satisfied that it is perfectly sound. This is not an opinion hastily adopted, but a conclusion based upon many observations, extending over a long period of years, and confirmed by experiments made upon myself.

"The *Pall Mall Gazette* of August 7th says: 'There is balm for tea-drinkers in one of Mr. Mattieu Williams's "Science Notes" in the *Gentleman's Magazine*.' This is true to a certain extent. I referred to the Chinese as habitual drinkers of boiled water, and suggest that this may explain their comparative immunity from cholera, where all the other conditions for a raging epidemic are fulfilled. It is the boiling of the water, not the infusion of tea-leaves therein, to which I attribute the destruction of the germs of infection.

"In the note which follows, I proposed an infusion of fried or toasted bread crumbs, oatmeal, maize, wheat, barley, malt, etc., as a substitute for the tea, the deep color of the infusion (poured off from the grounds in this case) serving to certify the boiling of the water. Rumford's burned soup, taken habitually at breakfast or other meals, would answer the same purpose, with the futher advantage to poor people of being, to a certain extent, a nutritious soup as well as a beverage. All that is nutritious in porter is in this, minus the alcoholic drug and its vile companion, the fusel-oil.

"The experience of every confirmed tea-drinker, when soundly interpreted, supplies condemnation of the beverage; the plea commonly and blindly urged on its behalf being, when understood, an eloquent expression of such condemnation. 'It is so refreshing;' 'I am fit for nothing when tea-time comes round until I have had my tea, and then I am fit for anything.' The 'fit-for-nothing' state comes on at five P. M., when the drug is taken at the orthodox time, or even in the early morning, in the case of those who are accustomed to have a cup of tea brought to their bedside before rising. With blindness still more profound, some will plead for tea by telling that by its aid one can sit up all night long at brain-work without feeling sleepy, provided ample supplies of the infusion are taken from time to time.

"It is unquestionably true that such may be done; that the tea-drinker is languid and weary at tea-time, whatever be the hour, and that the refreshment produced by 'the cup that cheers' and is *said* not to inebriate, is almost instantaneous.

"What is the true significance of these facts?

"The refreshment is certainly not due to nutrition, not to the rebuilding of any worn-out or exhausted organic tissue. The total quantity of material conveyed from the tea-leaves into the water is ridiculously too small for the performance of any such nutritive function; and, besides this, the action is far too rapid, there is not sufficient time for the conversion of even that minute quantity into organized working tissue. The action cannot be that of a food, but is purely and simply that of a stimulating or irritant drug, acting directly and abnormally on the nervous system.

"The five-o'clock lassitude and craving are neither more nor less than the reaction induced by the habitual abnormal stimulation; or otherwise, and quite fairly, stated, it is the outward symptom of a diseased condition of brain produced by the action of a drug; it may be but a mild form of disease, but it is truly a disease nevertheless.

"The active principle which produces this result is the crystalline alkaloid, the *theine*, a compound belonging to the same class as strychnine and a number of similar vegetable poisons. These, when diluted, act medicinally, that is, produce disturbance of normal functions as the tea does, and, like theine, most of them act specially on the nervous system; when concentrated they are dreadful poisons, very small doses producing death.

"The non-tea-drinker does not suffer any of these five-o'clock symptoms, and, if otherwise in sound health, remains in steady working condition until his day's work is ended and the time for rest and sleep arrives. But the habitual victim of any kind of drug or disturber of normal functions acquires a diseased condition, displayed by the loss of vitality or other deviation from normal condition, which is temporarily relieved by the usual dose of the drug, but only in such wise as to generate a renewed craving. I include in this general statement all the vice-drugs (to coin a general name), such as alcohol, opium, tobacco (whether smoked, chewed, or snuffed), arsenic, hashish, betel-nut, coca-leaf, thorn-apple, Siberian fungus, maté, etc., all of which are excessively 'refreshing' to their victims, and of which the use may be, and has been, defended by the same arguments as those used by the advocates of habitual tea-drinking.

"Speaking generally, the reaction or residual effect of these on the system is nearly the opposite of that of their immediate effect, and thus larger and larger doses are demanded to bring the system to its normal condition. The non-tea-drinker, or moderate drinker, is kept awake by a cup of tea or coffee taken late at night, while the hard drinker of these beverages scarcely feels any effect, especially if accustomed to take it at that time.

"The practice of taking tea or coffee by students, in order to work at night, is downright madness, especially when preparing for an examination. More than half of the cases of breakdown, loss of memory, fainting, etc., which occur during severe examinations, and far more frequently than is commonly known, are due to this.

"I frequently hear of promising students who have thus failed; and, on inquiry, have learned—in almost every instance —that the victim has previously drugged himself with tea or coffee. Sleep is the rest of the brain; to rob the hard-worked brain of its necessary rest is cerebral suicide.

"My old friend, the late Thomas Wright, was a victim of this terrible folly. He undertook the translation of the 'Life of

Julius Cæsar,' by Napoleon III., and to do it in a cruelly short time. He fulfilled his contract by sitting up several nights successively by the aid of strong tea or coffee (I forget which). I saw him shortly afterwards. In a few weeks he had aged alarmingly, and had become quite bald; his brain gave way and never recovered. There was but little difference between his age and mine, and but for this dreadful cerebral strain, rendered possible only by the alkaloid (for otherwise he would have fallen to sleep over his work, and thereby saved his life), he might still be amusing and instructing thousands of readers by fresh volumes of popularized archæological research.

"I need scarcely add that all I have said above applies to coffee as to tea, though not so seriously in this country [England]. The active alkaloid is the same in both, but tea contains, weight for weight, about three times as much as coffee. In this country we commonly use about fifty per cent. more coffee than tea to each given measure of water, and thus get about half as much alkaloid. On the Continent they use about double our quantity (this is the true secret of 'coffee as in France'), and thus produce as potent an infusion as our tea.

"The above remarks are exclusively applied to the *habitual* use of these stimulants. As medicines, used occasionally and judiciously, they are invaluable, provided always that they are not used as ordinary beverages. In Italy, Greece, and some parts of the East, it is customary, when anybody feels ill, with indefinite symptoms, to send to the druggist for a dose of tea. From what I have seen of its action on non-tea-drinkers, it appears to be specially potent in arresting the premonitory symptoms of fever, the fever-headache, etc.

"It is strange that any physiologist should claim this diminution of the normal waste and renewal of tissue as a merit, seeing that life itself is the product of such a change, and death the result of its cessation. But, in the eagerness that has been displayed to justify existing indulgences, this claim has been extensively made by men who ought to know better than admit such a plea.

"I speak, of course, of the *habitual* use of such drugs, not of their occasional medicinal use. The waste of the body may be going on with killing rapidity, as in fever, and then such medicines may save life, provided always that the body has not become 'tolerant' of or partially insensible to them by daily usage. I once watched a dangerous case of typhoid fever. Acting under the instructions of skilful medical attendants, and aided by a clinical thermometer and a seconds-watch, I so applied small doses of brandy at short intervals as to keep down both pulse and temperature within the limits of fatal combustion. The patient had scarcely tasted alcohol before this, and therefore it exerted its maximum efficacy. I was surprised at the certain response of both pulse and temperature to this most valuable medicine and most pernicious beverage.

"The argument that has been the most industriously urged in favor of all the vice-drugs, and each in its turn, is that miserable apology that has been made for every folly, every vice, every political abuse, every social crime (such as slavery, polygamy, etc.), when the time has arrived for reformation. I cannot condescend to seriously argue against it, but merely state the fact that the widely diffused practice of using some kind of stimulating drug has been claimed as a sufficient proof of the necessity or advantage of such practice. I leave my readers to bestow on such a plea the treatment they may think it deserves. Those who believe that a rational being should have rational grounds for his conduct will treat this customary refuge of blind conservatism as I do."

Mr. Williams, in his article, proceeds to give the views of certain scientists who have defended the use of the alkaloids. He speaks of Liebig's, or rather Nehmen's, theory, which was that the use of tea and coffee retarded the waste of the tissues of the body; also Johnston's theory, "Chemistry of Common Life," that if waste be lessened by the use of tea, less food is required.

Mr. Williams says, regarding these theories: "All the popular stimulants and refreshing drugs have two distinct and opposite actions; an immediate exaltation, which lasts for a certain period, varying with the drug and the constitution of its victim, and a subsequent depression proportionate to the primary exaltation, but, as I believe, always exceeding it either in duration or intensity, or both, thus giving as a net or mean result a loss of vitality."

Remarks on the Influence of Alcoholic Liquors, by Professor Edward L. Youmans, in " Household Science," and others.

"STIMULATING EFFECT OF ALCOHOLIC BEVERAGES.—They produce general stimulation; the heart's action is increased, the circulation quickened, the secretions augmented, the system glows with unusual warmth, and there is a general heightening of the functions. Organs usually below par from debility are brought up to the normal tone, while those which are strong and healthy are raised above it. Thus the stomach, if feeble, for example, from deficient gastric secretion, may be aided to pour out a more copious solvent, which promotes digestion; or, if it be in full health, it may thus be made to digest more than the body requires. The life of the system is exalted above its standard; which takes place, not by conferring additional vitality, but by plying the nervous system with a fiery irritant, which provokes the vital functions to a higher rate of action. This is the secret of the fatal fascination of alcohol, and the source of its evil. The excitement it produces is transient, and is followed by a corresponding depression and dragging of all the bodily movements. It enables us to live at an accelerated speed to-day, but it is only plundering to-morrow. By its means we crowd into a short period of intense exhilaration the feelings, emotions, thoughts, and experiences which the Author of our nature designed should be distributed more equally through the passing time. We cannot doubt that God has graduated the flow of these life-currents in accordance with the profoundest harmonies of being and the highest results of beneficence. By habitually resorting to this potent stimulant man violates the providential order of his constitution, loses the voluntary regulation and control of his conduct, inaugurates the reign of appetite and passion, and reaps the penal consequences in multiform suffering and sorrow—for Nature always vindicates herself at last."

Prof. Youmans also says, in answer to the question, Is the use of alcohol physiologically economical? "The apologists for the general and moderate use of alcoholic beverages cannot agree among themselves upon any philosophy to suit the case. Dr. Moleshott says, 'Alcohol may be considered a savings-box of the tissues. He who eats little, and drinks a moderate quantity of spirits, retains as much in the blood and tissues as a person who eats proportionally more without drinking any beer, wine, or spirits. Clearly, then, it is hard to rob the laborer, who, in the sweat of his brow, eats but a slender meal, of a means by which his deficient food is made to last him a longer time.' Upon which Dr. Chambers justly remarks, 'This is going rather too far. When alcohol limits the consumption of tissue, and so the requirements of the system, while at the same time a man goes on working, it is right to inquire, whence comes his new strength? It is supplied by something which is not decomposition of tissue; by what, then? Dr. Liebig points out the consequences of that peculiar economy by which the laboring man saves his tissue and the food necessary to repair it by the use of liquors: 'Spirits, by their action on the nerves, enable the laborer to make up for deficient power (from insufficient food) *at the expense of his body ;* to consume to-day that quantity which ought naturally to have been employed a day later. He draws, so to speak, a bill on his health which must be always renewed, because, for want of means, he cannot take it up; he consumes his capital instead of his interest, and the result is the inevitable bankruptcy of his body.'

"Dr. Moleshott further says, 'When, by habit, the stimulant has become a necessity, an enervating relaxation infallibly follows, as is sometimes mournfully illustrated by less prudent literary men. The stimulant ceases to excite; the debilitated organs have already been indebted to it for all the activity it can give. In this case the victim continues to seek his refuge until dangerous diseases of the stomach cripple the digestive organs, the formation of blood and nutrition are disturbed; and, with the digestion, vanish clearness of thought, acuteness of the senses, and the elasticity of the muscles.'"

Tendency of Common Wheat Flour to Produce Bright's Disease, Diabetes, etc.

It is claimed by the health-food manufacturers that "the starch portion of wheat may be compared to the fat of meat, and the gluten portion to the lean meat. This comparison is not wanting in scientific accuracy, inasmuch as starch is carbon and fat is carbon, while animal albumen and gluten, or vegetable albumen, are nearly identical nitrogenous substances. If, then, we were to attempt to exist upon the fat, or carbon, to the exclusion of the lean, or nitrogen, of meat, we should presently discern, by our waning bodily and mental vigor, that we were very imperfectly nourished. The same lack of vital force comes from an excessive use of the vegetable carbons. The disuse of the fat of grain—the starch—demands more earnest consideration from the physiologist, because the refined taste instinctively shrinks from the copious use of animal fats, while education, custom, habit, all encourage the increasing and unlimited use of the starch form of carbon. It is not claimed that our ordinary bread-flour is as pure a carbon, as free from nitrogen, as the clear fat of meat. The ordinary milling processes cannot exclude all the nitrogenous elements from the white flour; that they do withhold the greater part, as well as all but the merest trace of the organized mineral constituents, is a simple chemical fact. We know that the gluten contains phosphorus . . . we know that the starch contains no phosphorus. We know that the starch-interior of the wheat-berry is nearly barren of minerals, containing considerably less than one half of one per cent., while the gluten is found to contain over eleven per cent. The mineral matter is nearly half phosphoric acid, nearly one third potassa, more than one tenth magnesia, with smaller proportions of soda, lime, iron, chloride of sodium, sulphuric acid, and silica. These elements are all demanded in the blood-making processes. . . . In the use of starch-bread the stomach is greatly overtaxed in its effort to digest an immense amount of starch, containing an insignificant portion of nitrogenous and mineral elements. The use of starch in excess is the rule in America. If assimilated, it is very liable to induce fatty degeneration of the tissues, and such diseases as depend upon this state. Atheroma of the cerebral arteries, with the attendant fat-globules, the weakened muscular coats, and the tendency to rupture and apoplexy, are all concomitants of the starchy diathesis. The essential feature of Bright's disease is fatty infiltration of the kidneys; while diabetes finds its chief allies in bread and potatoes. These formidable diseases may be guarded against by appropriate alimentary substances containing the needed proportions of all nutritive elements.

But starch undigested is nearly as potent for evil as starch digested. The liver, burdened with white bread and potatoes, seems presently to be deprived of its power, etc.

Koumiss.

In the *Medical Record* is an article by Dr. E. F. Brush, of New York, in which he says: "Historically the study of koumiss is very interesting. Homer speaks of the koumiss-making Hippomolgi; Herodotus tells us that the Scythians deprived their slaves of sight in order to keep secret the process of making a drink from mares' milk. . . . Marco Polo, the great Venetian traveller, writing a few years later, speaks of koumiss as a common drink, wholesome, nutritious, and possessing important medical properties. . . . Pallas, who was sent by the Empress Catherine II. to visit the less-known portions of her dominions, gave considerable attention to the question of koumiss. Speaking of the Tartar tribes, he says: 'Their wealth consists in herds of mares, the milk of which cannot be manufactured into cheese or butter, and which, owing to the large quantity of sugar it contains, ferments spontaneously. This they undoubtedly discovered by attempting to preserve the milk for a day or two in skin bags. From this step, it is a short one to discover that the longer it was kept the more pleasant it became.' Mrs. Guthrie, who visited the Crimea in 1795, writes: 'On stopping at a village the hospitable Tartars brought us a wooden dish of their favorite koumiss. The koumiss has a sourish-sweet taste, by no means unpleasant to my palate.' Pallas tells us that he met a horde of Tartars who possessed the secret of turning cows' milk into vinous fermentation, or, in other words, into koumiss. Atkinson, in his 'Oriental and Western Siberia,' writes: 'On entering a Kirghis yourt in summer, a Chinese bowl holding three pints of koumiss is presented to each guest. It is considered impolite to return the vessel before emptying it, and a good Kirghis is never guilty of this impropriety. They begin to make koumiss in April. The mares are milked into large leathern pails, which are immediately taken into the yourt, and the milk poured into the koumiss bag. The first fourteen days after they begin making this beverage very little of it is drank, but, with fermentation and agitation, it is considered by this time in perfection, when it is drank in great quantities by the wealthy Kirghis.'

"In an official report to the Russian government in 1840, Dr. Dahl, after describing the method of manufacturing koumiss, continues: 'Peculiar as is the taste of koumiss, one soon becomes accustomed to it, especially if one tastes it for the first time when thirsty, or after violent exercise. It is then the most pleasant and refreshing of all drinks. . . . It is very refreshing and hunger-stilling, without being surfeiting. It only allays hunger without destroying the appetite. One can, without any fear, drink as much as he will—an inconceivable amount—and yet always feel light and well. If one were to drink half the quantity of water, beer, or anything else, especially during the burning heat when one is forced to be on horseback, one would feel over full and heavy. But every cup of koumiss gives new courage and strength. An intoxication such as is produced by wine never takes place after drinking koumiss, in whatever quantities you may; the result is a scarcely noticeable exhilaration, and this only when it is taken in very considerable quantitics, or in delicate persons, when it produces an inclination to a refreshing sleep. . . . Koumiss is, among the nomads, the drink of all children from the suckling upward, the refreshment of the old and sick, the nourishment and greatest luxury of every one. The effect of koumiss shows itself in less than a week in a good nourishment of the whole body, an increase in strength and spirits, and a general feeling of health. The respiration is easier, the voice freer, the complexion brighter. . . . The diseases in which koumiss is beneficial are those where the body must be well nourished without loading the digestive organs. It seems too, that koumiss is specially useful in diseases of the lungs, bronchia, and larynx; I will not assert that it can cure consumption and phthisis, but it suits these conditions better than any other nourishment. It is certain that among the Kirghis consumption and phthisis are very rare—so, too pneumonia, senile asthma, and dropsy of the chest. Of tubercular consumption, and other phthisis, I have seen no example among the Kirghis.'

"Dr. Neftel, who, twenty-three years after the visit of Dr. Dahl, was also sent by the Russian government to the Kirghis Steppe, confirms the observations of his predecessor. 'Scrofulosis and rachitis are quite unknown among them; and, what is still more remarkable, I had opportunity to observe not one single case of lung tuberculosis although I sought for such cases with great attention.' To avoid repetition, I will simply cite one case given by Dr. Neftel relating to koumiss treatment. 'The patient, twenty-five years old, had always lived in St. Petersburg. Her physician there, a distinguished diagnostician, found tubercular infiltrations in both superior lobes of the lungs. During two years she coughed continually, with a muco-prurient expectoration often tinged with blood, and she became very emaciated. All other physicians consulted by the patient confirmed this diagnosis. . . . The presence of cavities was clearly demonstrated, and a hectic fever set in. In this condition the patient, by my advice, left the city, passed the whole summer in the steppe, in a *kibitka*, and was methodically treated with koumiss. Her general condition gradually improved; she returned to the city in the autumn, and the ensuing spring she again commenced the koumiss treatment, and I have lately received here at Würzburg a letter from her husband, in which he informs me that his wife is completely cured, and coughs no longer.'"

Dr. Brush further adds that a recent article on koumiss has been written by Dr. Campbell, of Mount Vernon, N. Y., in the *American Journal of Obstetrics*, Oct., 1880. His observations are limited to the study of koumiss in cholera infantum. He reasons as follows: "In a severe case of choleraic diarrhœa we derive but little aid from medication, the primary cause of the disorder being the food put into the child's stomach. These cases occur almost exclusively among fed children. Our aim is chiefly directed to finding something on which the infant can be nourished and which will not increase the trouble already existing. In koumiss we have a food which children with high temperature not only take kindly, but crave, its slightly acid taste being grateful to their parched tongues. It is an absolutely non-putrefactive food, is free from sugar, and is rarely ejected even by the most irritable stomach. . . . I can say of it that it has never failed me in any case of cholera infantum, except where well-marked brain symptoms already existed, before it was administered, to such a degree as to preclude the possibility of a recovery. Even in these cases it is an advantage, for we are giving a food which will not be vomited, and which will satisfy thirst."

As a food for diabetics the author would refer to page 10.

Remarks by Dr. T. Griswold Comstock on the Use of Koumiss:
"Regarding koumiss, from a large experience in its use during the past nine years, I can recommend it with the greatest confidence. It fills a desideratum which the medical practitioner has long desired. One fact bearing upon its nutritious value should be borne in mind: *one pint of it contains more than two ounces of solid food*, so that it is especially indicated in constitutional diseases or systemic affections. According to the most recent authorities it is regarded by practitioners

as acting in cold weather as a diuretic, and in warm weather as a diaphoretic. From these physiological standpoints we can prescribe it rationally in a variety of ailments. It is valuable in pulmonary catarrh, in pulmonary tuberculosis, in chronic diarrhœa, in diabetes, in Bright's disease, in diphtheria, in the paralysis the sequel of diphtheria, in summer complaint, in the chronic intestinal and gastric catarrhs of children or adults, and especially in dyspepsia and flatulence. It will be found peculiarly beneficial in cases of incurable disease, such as cancer. I have prescribed it in pernicious anæmia, puerperal anæmia, in typhoid fever, in puerperal fever; in fact, in almost any affection attended with emaciation. At first it may be given in small quantities, and gradually the ration may be increased until it constitutes the sole food of the patient. As it is in reality a wine-milk, or rather a champagne-milk, it acts something like an alcoholic stimulant, and most patients feel revived at once after taking it. It is especially indicated for the infirmities of old age, in cases of palsy, paralysis, impending or real mental affections, etc.

From Dr. Roberts's Book, "The Digestive Ferments."

"My own efforts to produce a palatable peptonized food have been chiefly directed to the pancreatic method. The pancreas excels the stomach as a digestive organ, in that it has the power to digest the two great alimentary principles, starch and proteids; and an extract of the gland is possessed of similar properties. . . . My attention was first turned to the artificial digestion of milk. . . . Milk contains all the elements of a perfect food, adjusted in their due proportions for the nutrition of the body. Two out of three of its organic constituents—namely, the sugar and the fat—exist already in the most favorable condition for absorption, and require little, if any, assistance from the digestive ferments. It is therefore obvious that if we could change the caseine of milk into peptone without materially altering the flavor and appearance of the milk, such a result would go far towards solving the problem of supplying an artificially digested food for the use of the sick."

PEPTONIZED MILK GRUEL.—Dr. Roberts further says: "This is the preparation of which I have had the most experience, and with which I have obtained the most satisfactory results. It may be regarded as an artificially digested bread-and-milk, and as forming by itself a complete and highly nutritious food for weak digestion. . . . I find, however, that some persons fail to peptonize milk gruel so as to make it palatable. This is entirely due to allowing the peptonizing process to go on too far. Artificial digestion, like cooking, must be regulated as to its degree. If the *liquor pancreaticus* is very active, the slight bitterness, whereby it is known that the process has been carried far enough, is developed in an hour or less, but if the preparation is not so active, two or three hours may be required to reach the same point. The practical rule for guidance is to allow the process to go on until a perceptible bitterness is developed, and not longer. The milk gruel should be raised to the boiling-point to put a stop to further changes."

PANCREATIC EMULSION OF FATS.—Dr. Dobell, in his work on "Loss of Weight, Blood-spitting, and Lung Disease," says: "Oil when it agrees and passes into the blood does not completely represent the solid fats of the natural food, and cannot therefore permanently take their place. As a temporary substitute for natural fat it answers admirably, but sooner or later, in some cases very soon indeed, the portal system becomes choked and refuses to absorb more oil; the oil disagrees with the stomach, it rises, spoils the appetite, and thus not only ceases to do good, but does positive harm, by preventing the patient from taking as much food as the stomach might otherwise call for and digest. None of these disadvantages occur with well-made pancreatic emulsions of solid fat. The consequence is that an artificial supply of natural fat by the natural route can be kept up for an indefinite time if required, while the appetite is usually improved and the digestion also;

and at the same time a very large quantity of amylaceous* food is rapidly converted into dextrine and sugar by the pancreatic action of the emulsion, and thus a most important assistance in the economy of fat is given by the increased supply of carbon from the carbohydrates† at the same time that fat is being thrown into the blood by the emulsion.

"From the date of its first introduction in 1863 up to 1872, at the Royal Chest Hospital alone, I had prescribed the emulsion in over six thousand cases. . . . The general results of my thus extended experience have been confirmatory of my opinion. . . . I am informed on good authority that as much as sixty thousand pounds of the emulsion (made in London) have been consumed in a single year. While there are certainly a few persons who cannot possibly take or assimilate the emulsion, although able to take cod-liver oil, they are but very few indeed, now that the emulsion has been made so perfect a preparation; whereas the number of persons who can take and assimilate the emulsion but not cod-liver oil, is very large. In either case, it is necessary not to be too easily persuaded by our patients from prescribing the remedy. I frequently find that patients who assert that they cannot possibly, and never could, keep down the oil, will manage to do so when informed that it is the only thing that will stay the progress of the disease."

Food for Infants.

Remarks of Dr. Eustace Smith, Physician to the King of the Belgians, in *The Sanitary Record:*

"The mortality among children under the age of twelve months is enormous, and of these deaths a large proportion might be prevented by a wider diffusion of knowledge of one of the most simple of subjects. . . . The great principle at the bottom of all successful feeding, viz., that an infant is nourished in proportion to his power of digesting the food with which he is supplied, and not in proportion to the quantity of nutritive material which he may be induced to swallow, is so

---

* Pertaining to starch.   † Sugar and starch.

obviously true that an apology might almost seem necessary for stating so self-evident a proposition; but experience shows that this simple truth is one which, in practice, is constantly lost sight of. That that child thrives best who is most largely fed, and that the more solid the food the greater its nutritive power, are two articles of faith so firmly settled in the minds of many persons that it is very difficult indeed to persuade them to the contrary. To them wasting in an infant merely suggests a larger supply of more solid food; every cry means hunger, and must be quieted by an additional meal. To take a common case: A child, weakly, perhaps, to begin with, is filled with a quantity of solid food which he has no power of digesting. His stomach and bowels revolt against the burden imposed upon them, and endeavor to get rid of the offending matter by vomiting and diarrhœa; a gastro-intestinal catarrh is set up, which still further reduces the strength; every meal causes a return of the sickness; the bowels are filled with fermenting matter, which excites violent griping pains, so that the child rests neither night nor day; after a longer or shorter time he sinks, worn out by pain or exhaustion, and is then said to have died from 'consumption of the bowels.'

"Cases such as the above are but too common, and must be painfully familiar to every physician who has much experience of the diseases of children.

"The food we select for the diet of an infant should be nutritious in itself, but it should also be given in a form in which the child is capable of digesting it; otherwise we may fill him with food without in any way contributing to his nutrition, and actually starve the body while we load the stomach to repletion. No food can be considered suitable to the requirements of the infant unless it not only possess heat-giving and fat-producing properties, but also contains material to supply the waste of the nitrogenous tissues; therefore a merely starchy substance, such as arrowroot, which enters so largely into the diet of children, especially among the poor, is a very undesirable food for infants, unless given in very small quantities and mixed largely with milk.

"The most perfect food for children—the only one, indeed, which can be trusted to supply in itself all the necessary elements of nutrition in the most digestible form—is milk. In it are contained nitrogenous matter in the curd, fat in the cream, besides sugar, and the salts which are so essential to perfect nutrition. The milk of different animals varies to a certain extent in the proportion of the several constituents, some containing more curd, others more cream and sugar; but the milk of the cow, which is always readily obtainable, is the one to which recourse is usually had, and, when properly made, this is perfectly efficient for the purpose required. Cow's milk contains a larger proportion of curd and cream, but less sugar, than is found in human milk, and these differences can be immediately remedied by dilution with water and the addition of cane or milk sugar in sufficient quantity to supply the necessary sweetness. But there is another and more important difference between the two fluids which must not be lost sight of. If we take two children, the one fed on cow's milk and water, the other nursed at his mother's breast, and produce vomiting after a meal by friction over the abdomen, we notice a remarkable difference in the matters ejected. In the first case we see the curd of the milk coagulated into a firm, dense lump; while in the second the curd appears in the form of minute flocculent, loosely connected granules. The demand made upon the digestive powers in these two cases is very different, and the experiment explains the difficulty often experienced by infants in digesting cow's milk, however diluted it may be; for the addition of water alone will not hinder the firm clotting of the curd. In order to make such milk satisfactory as a food for new-born infants further preparation is required; and there are two ways in which the difficulty may be overcome.

"Although any thickening matter will have the mechanical effect desired of separating the particles of curd, yet it is not immaterial what substance is chosen. The question of the farinaceous feeding of infants is a very important one, for it is to an excess of this diet that so many of their derangements may often be attributed. Owing to a mistaken notion that such foods are peculiarly light and digestible—a notion so widely prevalent that the phrase "food for infants" has become almost synonymous with farinaceous matter—young babies are often fed as soon as they are born with large quantities of corn-flour or arrowroot, mixed sometimes with milk, but often with water alone. Now starch, of which all the farinas so largely consist, is digested principally by the saliva, aided by the secretion from the pancreas, which convert the starch into dextrine and grape-sugar previous to absorption. But the amount of saliva formed in the new-born infant is excessively scanty, and it is not until the fourth month that the secretion becomes fully established. Again, according to the experiments of Korowin, of St. Petersburg, the pancreatic juice is almost absent in a child of a month old; even in the second month its secretion is very limited, and has little action upon starch. It is only at the end of the third month that its action upon starch becomes sufficiently powerful to furnish material for a quantitative estimation of the sugar formed. Therefore, before the age of three months a farinaceous diet is not to be recommended—is even to be strongly deprecated, unless the starchy substance be given with great caution and in very small quantities. If administered recklessly, as it too often is, the food lies undigested in the bowels, ferments, and sets up a state of acid indigestion which, in so young and feeble a being, may lead to the most disastrous consequences. In fact, the deaths of so many children under two or three months old can be often attributed to no other cause than a purely functional abdominal derangement, excited and maintained by too liberal feeding with farinaceous foods. There is, however, one form of food which, although farinaceous, is yet well digested, even by young infants, if given in moderate quantities. This is barley water. The starch it contains is small in amount and is held in a state of very fine division. When barley water is mixed with milk in equal proportions it insures a fine separation of the curd, and is at the same time a harmless addition to the diet. Isinglass or gelatine, in the proportion of a teaspoonful to the bottleful of milk and water, may also be made use of, and will be found to answer the purpose well. Farinaceous foods in general are, as has been said, injurious to young babies on account of the deficiency during the first months of life of the secretions necessary for the conversion of the starch into the dextrine and grape-sugar, a preliminary process which is indispensable to absorption. If, however, we can make such an addition to the food as will insure the necessary chemical change, farinaceous matter ceases to be injurious. It has been found that, by adding to it malt in certain proportions, the same change is excited in the starch artificially as is produced naturally by the salivary and pancreatic secretions during the process of digestion. The employment of malt for this purpose was first suggested by Mialhe, in a paper read before the French Academy in 1845, and the suggestion was put into practice by Liebig, fifteen years later.

"'Liebig's Food for Infants' contains wheat flour, malt, and a little carbonate of potash, and has gained a well-deserved celebrity as a food for babies during the first few months of life. The best form with which I am acquainted is that made by Mr. Mellin, under the name of 'Mellin's Extract for Preparing Liebig's Food for Infants.' In this preparation, owing to the careful way in which it is manufactured, the whole of the starch is converted into dextrine and grape-sugar, so that the greater part of the work of digestion is performed before the food reaches the stomach of the child. Mixed with equal parts of milk and water this food is as perfect a substitute for mother's milk as can be produced, and is readily digested by the youngest infants. It very rarely, indeed, happens that it is found to disagree.

"In all cases, then, where a child is brought up by hand, milk should enter largely into his diet; and during the first few months of life he should be fed upon it almost entirely. If he can digest plain milk and water, there is no reason for making any other addition than that of a little milk, sugar, and cream; but in cases where, as often happens, the heavy curd taxes the gastric powers too severely, the milk may be thickened by an equal proportion of thin barley water, or by adding to each bottleful of milk and water a teaspoonful of isinglass or of Mellin's Extract."

ONE MONTH.

"Having fixed upon the kind of food which is suitable to the child, we must next be careful that it is not given in too large quantities, or that the meals are not repeated too frequently. If the stomach be kept constantly overloaded, even with a digestible diet, the effect is almost as injurious as if the child were fed upon a less digestible food in more reasonable quantities. A healthy infant passes the greater part of his time asleep, waking at intervals to take nourishment. These intervals must not be allowed to be too short, and it is a great mistake to accustom the child to take food whenever it cries. From three to four ounces of liquid will be a sufficient quantity during the first six weeks of life; and of this only a half or even a third part should consist of milk, according to the child's powers of digestion. After such a meal the infant should sleep quietly for at least two hours. Fretfulness and irritability in a very young baby almost always indicate indigestion and flatulence; and if a child cries and whines uneasily, twisting about its body and jerking its limbs, a fresh meal given instantly, although it may quiet it for the moment, will, after a short time, only increase the child's discomfort."

TWO MONTHS.

"During the first six weeks or two months, two hours will be a sufficient interval between the meals; afterwards this interval can be lengthened, and at the same time a larger quantity may be given at each time of feeding. No more food should be prepared at once than is required for the particular meal. The position of the child as it takes food should be half reclining, as when taking food from the mother's breast, and the food should be given from a feeding-bottle. When the contents of the bottle are exhausted the child should not be allowed to continue sucking at an empty vessel, as by this means air is swallowed which might afterwards be a source of great discomfort."

SIX MONTHS.

"At the age of six months farinaceous food may be given in small quantities with safety, if it be desired to do so; and in some cases the addition of a small proportion of wheaten flour to the diet is found to be attended with advantage. The best form in which this can be given is the preparation of wheat known as 'Chapman's Entire Wheaten Flour.' This is superior for the purpose to the ordinary flour, as it contains the inner husk of the wheat finely ground, and is, therefore, rich in phosphates and in a peculiar body called cerealine, which has the diastatic property of changing starchy matters into dextrine."

EIGHT MONTHS.

"After the eighth month a little thin mutton or chicken broth or veal tea may be given, carefully freed from all grease. After

TWELVE MONTHS

The child may begin to take light puddings, well-mashed potatoes with gravy, or the lightly boiled yolk of an egg; but no meat should be allowed until the child be at least sixteen months old. Every new article of food should be given cautiously and in small quantities at first; and any sign of indigestion should be noted, and a return be made at once to a simpler method of feeding."

Feeding the Baby.

Dr. C. E. Page, in a very admirable little book, "How to Feed the Baby," thinks babies are generally overfed. He thinks three meals a day and nothing at night, for an infant from its birth, is quite enough; that the stomach of an infant needs rest like that of an adult; that the stomach should be allowed to clear itself and rest before the next meal is taken; that "the stomach is generally forced to go to work again too soon, and later this excessive labor exhausts the muscular power of the stomach; the supply of gastric juice is not enough to digest unneeded food, which, if not thrown up, remains to putrefy and poison the blood." Dr. Page relates his experience with his own children (also others under his charge), who were brought up on the three-meals-a-day plan. He says they slept all night like older people. At the same time due attention was paid to ventilation. A little dropping of the upper window always kept the room well aired; no swaddling clothes pinched the vital organs.

He says: "If the child be fed and dressed properly, and is otherwise rationally managed, there will be no midnight orgies, no sleepless nights on baby's account, and it will soon, indeed in a very few days, become so regular in habit that the bundled, pinned-up squares, so sweltering and injurious, can be entirely dispensed with at night, and during its naps by day, and it may be safely laid down after supper for its ten or twelve hours of solid sleep."

What Dr. Page considers a sufficient amount of diet is as follows: "No definite rule can be given for the amount of food necessary for a hand-fed babe at any given age. It will not, however, vary much from one pint for an infant of six months. This amount, divided into three meals at 6 A.M., 12 M. and 6 P.M., has, in my experience, always insured the best results."

This seems very little, yet undoubtedly babies are generally overfed.

He also says: "During hot weather the child does not need as much food as in winter. . . . The baby should be allowed water frequently in summer."

Dr. Dawson, of New York, discussing the same subject, says: "When treating vomiting, constipation, or diarrhœa in children, the stomach is given rest by cutting off all but a small quantity of food. Will we gain any benefit, I ask, from ejected or undigested food, even if it causes no severer disturbance?"

Again he says: "Constipation, too, so common in otherwise healthy infants, is generally due to excessive and too-frequent feeding. The explanation is quite simple. The stomach being overburdened with food, and consequently overtaxed with work, each supply of milk, instead of being coagulated into fine and soft coagula, which are readily acted upon by the secreted pepsin, comes into contact with the semi-digested acid coagula of the preceding meal, and, in consequence, is coagulated more rapidly than it should be normally, the coagula being larger and harder. Such masses, if not ejected, pass into the intestinal canal but little or not at all changed by the digestive process, will impact together on contact, and from their size and dryness are with difficulty passed along the bowels, thus giving rise to constipation, colic, etc."

Professor Huxley says: "But, whatever the circumstances, if the quantity of food taken exceeds the demands of the system, evil consequences are sure to follow. The immediate results of overeating are lethargy, heaviness, and tendency to sleep. Overtaxing the digestive organs soon deranges their functions, and is a common and efficient cause of dyspepsia. If the food is not absorbed from the digestive apparatus into the system, it rapidly undergoes chemical decomposition in the alimentary canal, and often putrefies. Large quantities of gas are thus generated, which give rise to flatulence and colicky pains. Dyspepsia, constipation, and intestinal irritation causing diarrhœa are produced. If digestion be strong, and its products are absorbed, an excess of nutriment is thrown into the blood, and the circulation is overloaded. If food is not expended in force, the natural alternative is its accumulation in the system, producing plethora, and abnormal increase of tissue. This is accompanied by congestion of important organs, mal-assimilation of nutritive material, and increased proneness to derangement and diseased action."

Dr. Dawson says: "The ejection of milk after nursing, which is ignorantly considered by many to be the sign of a healthy child, denotes overfeeding, and is the effect of reflex action. . . . As my experience has taught me, most infants who thus throw up after eating suffer sooner or later from enteralgia and constipation, and other symptoms of indigestion, which later are only relieved when the greed of the child is restricted."

Dr. Page says: "One cause of excessive feeding exists in the desire of parents to have a *fat* baby. . . . The excessive fat, so generally regarded as a sign of a healthy babe, is as truly a state of actual disease as when it occurs at adult age. Not only are the muscles enveloped with fat, they are mixed with it throughout, and so are the vital organs—the kidneys, liver, heart, etc. Dissection in these cases often discloses the fact that these organs are enlarged and degenerated with fat; the liver, for example, is often double the normal size. The disease finally culminates in one of two things—a considerable period of nongrowth, or a violent sickness, which strips them of fat, if not of life."

Dr. Page further says: "It is not the large quantity swallowed, but the *right* quantity, properly digested and perfectly assimilated, that alone can insure the best results with either children or adults."

Diet for Typhoid Fever.

Extract from an address on the "Treatment of Typhoid Fever," delivered before the Midland Medical Society, 1879, by Sir William Jenner:

"From the first they should be restricted to a liquid diet with farinaceous food and bread in fine form, if the appetite should require it. It is better to vary the broths, and to add to them some strong essence of vegetables. Sometimes a little strained fruit juice is taken with advantage, but skins and seeds of fruits and particles of the pulp are frequent sources of irritation to the bowels. Grapes are always dangerous, from the difficulty of preventing seeds slipping down the throat. The value of milk as an article of diet is generally admitted, but it requires to be given with caution. The indiscriminate employment of milk in almost unlimited quantities as diet in fever has led to serious troubles. Milk contains a large amount of solid animal food. The caseine of the milk has to pass into a solid form before digestion can take place. Curds form in the stomach. Patients suffering from typhoid fever should be allowed an unlimited supply of pure water. When pure water is freely absorbed it passes away by the kidneys, skin, lungs, etc., and is of much service as a depurating agent. If it be possible even that the poison of the fever was conveyed into the patient by the drinking-water or the milk of the district in which he is ill, then these fluids should be boiled until a different supply is obtained. . . . The fever is thus met by rest, quiet, fresh air, mixed liquid food, and bland diluents, and the exclusion of fresh doses of poison; the intestinal lesion by careful exclusion from the diet of all hard and irritating substances, and the removal from the bowels of any local irritant.

"The chief causes of diarrhœa in excess of that due to the intestinal changes in typhoid fever are, first, errors in diet; second, the use of solid food—the presence of undigested food in the bowels, the abuse of milk and animal broths. My own experience has not satisfied me that one animal broth is more prone to produce diarrhœa than another. Excess of fluid, when there is irritability to absorb the quantity drank, passes through the bowels, and so stimulates excessive secretion from the intestinal mucous membrane.

"Alcohol in fit doses improves the nerve energy. . . . When blood in ever so small a quantity is observed in the secretions, the patient is to be kept in a recumbent position. He should not be allowed to make any effort whatever. All movement of the bowels should be restrained as far as possible and for as long as possible. . . . It is a point of the greatest moment to keep the bowels empty, and therefore nourishment should be given in the most concentrated and absorbable form; *i.e.*, essence of meat in tablespoonful doses, frequently repeated. Lumps of ice should be sucked, and all essence of meat iced.

"In a disease which runs a limited course, like typhoid fever, the greatest possible care should be taken to preserve the powers of the stomach, as the life of the patient may depend on his power to digest nourishment towards the end of his disease. . . . To avert death from failure of heart power alcohol is the great remedy. Over defective cardiac action— due altogether to changes in the muscular tissue, when once established, or in the circulation of poisoned blood through its vessels—alcohol exerts comparatively little influence; but when the weakness and frequency of cardiac action are due to nerve influence, in part or altogether, then alcohol exerts a singularly beneficial effect on the rapidity and feebleness of the heart's action. . . . I may sum up my experience in regard to the use of alcohol in the treatment of typhoid fever thus: Its influence is exerted primarily in the nervous system, and through it on the several organs and processes; for example, the heart and the general nutritive processes—changes on which the rise and fall of temperature depends. In judiciously selected cases it lowers temperature, increases the force and diminishes the frequency of the heart-beats; it calms and soothes the patient, diminishes the tremor; it quiets delirium, and induces sleep. It should never be given in the early stage of the disease, or with the hope of anticipating and so preventing the occurrence of prostration and debility, but should be prescribed only when the severity of special symptoms, or the general state of prostration, indicates its use. Hence a large number of cases of typhoid fever end favorably without alcohol being prescribed at all. It should not be prescribed when a sudden gush of blood has induced faintness, unless the faintness is so great as to threaten life immediately. Nor should it be given when, after the first few drops, the temperature rises, the heart's action becomes more frequent, or more feeble, delirium increases, sleeplessness supervenes, or drowsiness deepens, so as to threaten to pass into coma. When the urine contains a certain amount of albumen alcohol should not be prescribed unless absolutely necessary for the relief of some symptom immediately threatening life, and then it should be given with the greatest caution, and its effects on temperature and the circulation be carefully and frequently noted. The quantity of alcohol prescribed should be as much only as may be necessary to effect the object for which it is prescribed. In the fourth week, to tide the patient over the concluding days of the disease, it may, as a rule, be given more freely than in the second, or the beginning of the third, week of the disease; but it is in exceptional cases only. that more than twelve ounces of brandy in the twenty-four hours can be taken without inducing the worst symptoms of prostration. Nearly all the good effects of alcohol, when its use is indicated, are obtained by four, six, or eight ounces of brandy in twenty-four hours. Taken in excess, even when in smaller quantities, it would do the patient no good; it dries the tongue, muddles the mind. . . . When there is a question of a larger or a smaller dose, I, as a rule, give the smaller. The reverse of the rule I laid down for myself in the treatment of typhus fever."

Fresh Air and Diet for Colds and Catarrhs.

Extracts from "The Remedies of Nature," by Dr. Felix L. Oswald:

Dr. Oswald says: "That colds or catarrhal affections are so very common—more frequent than all other diseases taken together—is mainly due to the fact that the cause of no other disorder of the human organism is so generally misunderstood . . . the cause is taken for a cure, and the most effective cure for the cause of the disease. If we inquire after that cause, ninety-nine patients out of a hundred . . . would answer, 'Cold weather,' 'Raw March winds,' . . . in other words, out-door air of a low temperature. If we inquire after the best cure, the answer would be, 'Warmth and protection against cold draughts,' *i. e.*, warm, stagnant, in-door air. Now, I maintain that it can be proved . . . that warm, vitiated in-door air is the cause, and cold out-door air the best cure for catarrhs. . . . In all the civilized countries of the colder latitudes catarrhs are frequent in winter and early spring, and less frequent in midwinter, hence the inference. . . . No kind of warm weather will mitigate a catarrh while the patient persists in doing what thousands never cease to do the year round—namely, to expose their lungs night after night to the vitiated, sickening atmosphere of an unventilated bedroom. Colds are, indeed, less frequent in midwinter than at the beginning of spring. Frost is such a powerful disinfectant that in very cold nights the lung-poisoning atmosphere of few houses can resist its purifying influence; in spite of padded doors, weather-strips, and double windows, it reduces the in-door temperature enough to paralyze the floating disease germs. . . . All Arctic travellers agree that among the natives of Iceland, Greenland, and Labrador pulmonary diseases are actually unknown. Protracted cold weather thus prevents epidemic catarrhs, but during the first thaw nature succumbs to art, . . . the incubatory influence of the first moist heat is brought to bear on the lethargized catarrh germs. . . . Smouldering stove fires add their fumes to the effluvia of the dormitory; superstition triumphs; the lung-poison operates, and the next morning a snuffling, coughing, and red-nosed family discuss the cause of their affliction. . . . The summer season brings relief; . . . the windows are partially opened. The long warm days offer increased opportunities for out-door rambles. . . . No man can freeze himself into a catarrh. In cold weather the hospitals of our Northern cities sometimes receive patients with both feet and both hands frozen, . . . but without a trace of catarrhal affection. Duck hunters may wade all day in a frozen swamp without affecting the functions of their respiratory organs. Ice cutters not rarely come in for an involuntary plunge bath, and are obliged to let their clothes dry on their backs; it may result in a bowel complaint, but no catarrh. . . . Cold is a tonic that invigorates the respiratory organs when all other stimulants fail, and, combined with arm exercise and certain dietetic alteratives, fresh cold air is the best remedy for all the disorders of the lungs and upper air passages. . . . If the fight is to be strong and decisive (for breaking up a cold), the resources of the adversary must be diminished by a strict fast. . . . But, aided by exercise, out-door air of any temperature will accomplish the same result. In two days a resolute pedestrian can *walk away* from a summer catarrh of that malignant type that is apt to defy half-open windows. But the specific of the movement cure is *arm exercise*—a dumb-bell swinging, grapple-swing practice, and wood chopping. On a cold morning (for, after all, there are ten winter catarrhs to one in summer), a woodshed *matinée* seems to reach the seat of disease by an air line. As the chest begins to heave under the stimulus of the exercise, respiration becomes freer as it becomes deeper and fuller . . . mucus is discharged *en masse*, as if the system had only waited for that amount of encouragement to rid itself of the incubus. A catarrh can thus be broken up in a single day. For the next half week the diet should be frugal and cooling. Fruit, light bread (?), and a little milk, is the best catarrh diet." "A fast-day is still better. Fasting effects in a perfectly safe way what the old-school practitioners tried to accomplish by bleeding; it reduces the semi-febrile condition which accompanies every severe cold. There is no doubt but that by exercise alone a catarrh can gradually be 'worked off.' . . . A combination of the three specifics, exercise, abstinence, and fresh air, will cure the most obstinate cold."

This admirable article of Dr. Oswald's, published in the *Popular Science Monthly*, has undoubtedly done much to shake what he calls "the night-air superstition." Dr. Oswald sleeps with window wide open the year round, and he never has a cold. It would undoubtedly be indiscreet, however, to change a habit too suddenly.

The old maxim of "feeding a cold and starving a fever" is also refuted by physicians in general, as well as by Dr. Oswald.

Something more about the Pancreatic Extract for Artificial Digestion.

In the receipts I have given for the digestion of certain foods (pages 41 and 42), Dr. Benjamin T. Fairchild (the inventor of the "Pancreatic Extract" as prepared by Fairchild Brothers) tells me that he fears I allow too long a time for the digestive process, which renders the food less palatable. It is more satisfactory, he says, to digest the milk food but half an hour. If not taken immediately by the patient, the food is, after the half hour, placed on ice. This arrests digestion, and when the patient takes the food into the stomach, the digestion is there completed. As it is desirable to give the food to the patient warm, it can be slightly heated (a little more than lukewarm) just before it is administered. The boiling of the food kills the digestive principle of the extract used. Yet it is sometimes, in the absence of ice, desirable to boil it in order to keep it. The digestive function is not destroyed by cold temperature—only arrested.

I do not understand why it would not be as satisfactory to mix the pancreatic extract with the food just before eating, and allow the entire digestive process to be carried on in the stomach. I merely give the advice of others who ought to know more about it than the author.

I will also add that pancreatized barley gruel (made without sugar) is the most palatable of the pancreatized gruels.

I am also tempted to add a new receipt for a pancreatized food which is now much used.

Pancreatized Oysters.

Chop half a dozen raw oysters fine as possible, also pound them.

Bring two cupfuls (one pint) of the oyster liquor (it may be part water if there is not oyster liquor enough) to a boil, then thicken it with half a cupful of barley flour, rubbed smooth with half a cupful of water. Let it boil three or four minutes to cook the barley, then add the oyster pulp, and a seasoning of salt and very little pepper. When it comes to the boiling-point again, take it from the fire, and when the temperature is reduced to blood heat, mix in a fourth of a teaspoonful of pancreatic powder, and half a saltspoonful of soda. Pour it into a glass jar or bottle, and put this into water so hot that the whole hand can be held in without discomfort for a minute. Let it stand an hour as described for milk. It takes a little longer to digest oysters than milk.

The dish is most palatable served immediately. It is liable to curdle when brought to the boiling-point again. It can either be placed on ice, or brought to the boiling-point for the purpose of keeping.

ALPHABETICAL INDEX.

	PAGE
Animal Foods, remarks about	14
Apple Sauce	177
Apples, Baked	176
Arrow-root, remarks about	21
Asparagus Soup	161
" " (à la crême)	161
Babies, diet for	49
Barley Gruel	107
" Pudding	193
" Wafers	122
Bass à l'Espagnole	150
Beans, remarks about	21
Beef " "	14
" Juice	102
" Sandwich	144
Beef Tea	100
" " for Convalescents	103
" " for Invalids (Dr. Rice)	54
" " for Travelling	102
" " (Liebig's)	101
Beefsteak	141
" Chopped	143
Beets (à la crême)	136
Beverages from Fruits	92
Bird, a	149
Biscuits, Dixie	120
" Wafer	122
Blanc-mange, Corn-starch	178
" " Sea-moss	178
Bouillon	165
Bread	113
" Adirondack	114
" Boston Brown	116
" Corn	122
" " (No. 2)	123
" " (No. 3)	123
" " Rice	124
" Dice	103
" Gluten	130

	PAGE
Bread, Graham (No. 1)	115
" " (No. 2)	115
" Pulled	119
Bright's Disease, diet in	69
Broth, Beef, with a Poached Egg	106
" Chicken	104
" Clam	166
" Clear Beef	105
" Mutton	104
Buttermilk, remarks about	13
Cake, Coffee	120
" Graham Sponge	193
" Hoe	124
Carrots (à la crême)	136
Caudle, Oatmeal	109
Cauliflower	136
Charlotte-russe	179
Chicken, Breast of	144
" Broth	104
" Croquettes	145
" Fricassee	145
" Plain Boiled	147
" Prairie	149
" Soufflé	148
" Spring	147
" with Macaroni or Rice	146
Chocolate	98
" remarks about	5
Cholera, diet in	62
Chop, Mutton	144
Clabbered Milk	174
Clam Broth	166
" Soup	167
Coffee, remarks about	5
" to make	97
Compotes	177
Consumption, diet in	173
Corn Bread (No. 1)	122
" " (No. 2)	123

	PAGE
Corn Bread (No. 3)	123
" Rice Bread	124
" Soup	163
" " (No. 2)	164
Cornmeal Gruel	111
" Mush	129
" Pancakes	124
Corn-starch Blanc-mange	178
" " remarks about	21
Corpulent, diet for the	77
Cottage Cheese	175
Cracked Wheat	126
Crackers	58
Cream, a glass of	96
" Ice	175
" of Asparagus	161
" of Chicken	160
" of Corn	163
" " (No. 2)	164
" of Oysters	159
" of Potatoes	161
" of Rice, Farina, or Barley	160
" of String-beans	163
" Toast	118
" Whipped	174
Croquettes, Chicken	145
" Macaroni	154
" Oyster	156
Currant Jelly	185
" Preserves	186
" Scone	125
Custard à la Morrison	180
" Caramel	183
" Chocolate	182
" Coffee	181
" of Granulated Barley or Oats	182
" Plain Boiled	180
" Rennet	182
" Tapioca or Sago	181
Diabetes, diet in	71
Diarrhœa, "	61
Digestion, artificial	39
Diphtheria, diet in	76
Distilled Water	89
Dysentery, diet in	62
Dyspepsia "	55
Egg and Milk Punch	95
" Cordial	95

	PAGE
Eggnog	95
Eggs, Boiled	139
" Poached	139
" Raw	141
" remarks about	20
Farina Gruel	111
" Pudding	190
Fat, remarks about	18
Fever, diet in	64
" Typhoid	66, 222
Fish, Bass à l'Espagnole	150
" Boiled	150
" Broiled	149
" remarks about	16
Flaxseed and Licorice Tea	91
" Lemonade	94
" Tea	91
Flour Gruel	109
" " (No. 2)	110
" " (No. 3)	110
" Soup	167
Fruits, beverages from	92
" remarks about	22
" Stewed	177
Gastritis, diet in	76
Gelatine, remarks about	21
Gluten and Rice Muffins	131
" Bread	130
" Cheese Cakes	132
" Cream Wafers	132
" Muffins	131
" Mush	130
" Pudding or Gruel	131
" Soufflé	132
" Wafers	132
Gout, diet in	67
Graham Bread	115
" " (No. 2)	115
" Rolls	125
Graham-flour Pudding	190
Granulated-wheat Custard	182
" " Pancakes	124
" " Pudding	191
Grape Juice	44, 93
Grapes, remarks about	22
Gruel, Barley	107
" Cornmeal	111
" Farina	111
" Flour	109

Alphabetical Index.

	PAGE
Gruel, Flour (No. 2)	110
" " (No. 3)	110
" Graham-flour	108
" Oatmeal, for Infants	54
" " (No. 1)	108
" " (No. 2)	109
" Peptonized Milk	42
" Rice	110
Health-foods	26
Hot-water Cure	47
Ice-cream and Iced Peaches	175
Iced water, effects of	6
Infants, diet for	49, 214, 221
Jelly, Coffee	185
" Currant	185
" " Sauce	173
" " Water	91
" Peptonized Milk	42
" Wine	184
" " (without cooking)	185
Juice, Beef	102
" Grape	44
Koumiss	81
" to make	84
Lemon Pie or Pudding	193
Lemonade	93
" Flaxseed	94
Lime-water	89
Liquors, remarks about	8
Longevity	80
Macaroni and Tomato Sauce	152
" au Gratin	153
" Croquettes	154
" Pudding	191
" Soup	165
Malt Extract	9
Mellin's Food for Infants	54
Milk and Egg Punch	95
" and Seltzer-water	99
" Clabbered	174
" for Infants, Liebig's receipt	52
" Punch	94
" remarks about	10
" Toast	119
" to Peptonize	41

	PAGE
Mush, Cornmeal	129
" Gluten	130
Mutton Chop	144
" remarks about	14
Oatmeal Caudle	109
" Drink	90
" Gruel	54, 108, 109
" Porridge	128
" Wafers	122
Orange Marmalade	186
" Pudding	193
Oyster Croquettes	156
" Soup	166
Oysters (à la crème)	159
" on Toast	153
" remarks about	16
Oysters, Pancreatized	229
Panada	111
Pancakes, Cornmeal	124
" Flour	124
" Granulated-wheat	124
Pap	52
Pear Compote	177
Peppers, Stuffed	138
Potato Soup (à la crème)	161
Potatoes (à la crème)	135
" " (au gratin)	135
" to bake	134
" to boil	134
Preserves, Currant	186
" Strawberry	187
Pudding, Barley	193
" Corn Cottage	189
" Farina	190
" Graham Flour	190
" Granulated-wheat	191
" Lemon	193
" Macaroni	191
" Orange	193
" other Soufflée	192
" Quioque	191
Rennet Custard	182
Rheumatism, diet in	67
Rice à l'Imperatrice	172
" and Gravy	170
" Cones	170
" Dish, as a vegetable	173
" Gruel	110

Alphabetical Index.

	PAGE
Rice Pudding	170
" " (No. 2)	171
" " à la Guillod	172
" remarks about	20
" Soup (à la crême)	160
" to boil	169
" " in Milk	169
Rickets, diet in	75
Sago, remarks about	21
Salad	156
Sauce à la Guillod	192
" a plain one for Puddings	189
" Apple	177
" Brown	137
" Burke	192
" Currant Jelly	173
" Tomato	155
Scrofula, diet in	75
Sea-moss Blanc-mange	178
Seltzer-water and Milk	99
Spinach	135
Spirituous Liquors, remarks about	8
Soup, Asparagus	161
" " (à la crême)	161
" Chicken "	160
" Clam	167
" Corn (à la crême)	163
" " " (No. 2)	164
" Farina or Barley (à la crême)	160
" Flour	167
" Oyster	166
" " (à la crême)	159
" Potato "	161
" Rice "	160
" Stock for	164
" String-bean	163
Steak, Beef	141
" Venison	144
Strawberry Preserves	187
Sugar Syrup	92
Sweetbreads	151
" with Cream Sauce	152

	PAGE
Tamarind Water	90
Tapioca, Custard	181
" remarks about	21
Tea, Beef	100
" " (Liebig's)	101
" Flaxseed	91
" Iced	6
" remarks about	1, 198
" to make	96
Teas, Herb	91
Toast, Cream	118
" Milk	119
" Mock Cream	118
" Sippets	117
" to make	116
" Water	118
Tom and Jerry	95
Tomato Sauce	155
Tomatoes, remarks about	22
" Stuffed	137
Utensils	85
Venison Steak	144
Wafers, Oatmeal, Granulated-wheat, Barley, Gluten, etc.	122
Water, Apple	93
" Barley	90
" Cinnamon	90
" Currant-jelly	91
" Distilled	89
" Ice, remarks about	6
" Lime	89
" Oatmeal	90
" Tamarind	90
" Toast	91
Wheat, Cracked	126
Whey	92
" remarks about	13
Zwieback	119

THE END.

www.ingramcontent.com/pod-product-compliance
Lightning Source LLC
Chambersburg PA
CBHW020808230426
43666CB00007B/916